Benjamin Busch

Impact of grain coats on quartz overgrowth and reservoir properties

Impact of grain coats on quartz overgrowth and reservoir properties

by
Benjamin Busch

KIT Scientific Publishing

Dissertation, Karlsruher Institut für Technologie
KIT-Fakultät für Bauingenieur-, Geo- und Umweltwissenschaften

Tag der mündlichen Prüfung: 22. Dezember 2016
Erster Gutachter: Prof. Dr. Christoph Hilgers
Zweiter Gutachter: Prof. Dr. Helge Stanjek

Impressum

KIT Scientific Publishing

Karlsruher Institut für Technologie (KIT)
KIT Scientific Publishing
Straße am Forum 2
D-76131 Karlsruhe
KIT Scientific Publishing is a registered trademark
of Karlsruhe Institute of Technology.
Reprint using the book cover is not allowed.
www.ksp.kit.edu

Print on Demand 2017 – Gedruckt auf FSC-zertifiziertem Papier

ISBN 978-3-7315-0666-9
DOI 10.5445/KSP/1000069336

Impact of grain coats on quartz overgrowth and reservoir properties

zur Erlangung des akademischen Grades eines

Doktors der Naturwissenschaften

von der Fakultät für Bauingenieur-, Geo- und Umweltwissenschaften
des Karlsruher Instituts für Technologie (KIT)

genehmigte

Dissertation

von

Benjamin Busch

aus Viersen

Tag der mündlichen Prüfung: 22. Dezember 2016
Erster Gutachter: Prof. Dr. Christoph Hilgers
Zweiter Gutachter: Prof. Dr. Helge Stanjek

Acknowledgements

My sincerest thanks go to my advisor, Christoph Hilgers, his support, guidance, and continuous supply of new ideas were an essential part of this study. He also initiated the project, which is the basis of this work. Countless long discussions, especially shortly before submission deadlines, always lead to new and improved results (random topical excursions included).

Helge Stanjek is greatly acknowledged for being my second referee. Discussions and cooperation with him and his team greatly benefited parts of this work.

Dirk Adelmann is thankfully acknowledged for his countless hours reviewing written manuscripts, introducing me to Wintershall workflows and staff, discussions on diagenesis and samples, and helping me find my way around the Wintershall infrastructure in both German Wintershall sites.

The additional support by Philipp Antrett and discussions with and reviews by Michael Peter Süss, Norbert Schleifer, Petra David, Susan Weniger, Bernd Leonhardt, Bastian Koehrer, Stefano Dellepiane, Frank Jacobsen (all Wintershall Holding GmbH), Fred Kluin, Bert de-Wijn, Richard Huis in t Veld, and Arthur Rosenthal (all Wintershall Noordzee B.V.) on various topics concerning samples, basin modeling, additional methods or the general progress of the study are greatly appreciated. The financial support of this study by Wintershall Holding GmbH is thankfully acknowledged.

Rob Lander's and Linda Bonnell's contribution towards this project is highly appreciated. The donation of their Geocosm Touchstone™ software for academic purposes, their ongoing availability for discussions, feedback and polishing of manuscripts, conversations or a beer and meal in Durango are all thankfully acknowledged.

Thanks to Patrick for discussions on different topics, and generally not being the last one to leave the office in the evening. Furthermore, the discussions and input of Ulrike Hilse improved many ideas and manuscripts.

My colleagues Ivy Becker, Christina Schmidt, and Ernst Reichelt were always open for discussions and first-pass checks of ideas and models. My sincerest thanks to the students whose final theses were supervised by me. Bob Bamberg and Leonard Kaiser are thanked for excellent mapping results produced during their Bachelor thesis, supplying me with details on outcrops in Northern England. Christina Schmidt analyzed samples from eolian deposits in Germany and Northern England and contributed valuable data and results used in this project. Additionally, her samples gave us a first impression on flow heterogeneities around eolian bounding surfaces. Simon Schröer analyzed and modeled reservoir quality in the Permian Rotliegend. His results are integrated with the results of this study. Additionally, I would like to thank all Hiwis I was working with, be it Sebastian in Scotland, Dieter, Ivy and Tina in Southern England, or the many Pola and Petro Hiwis, during lectures and exams. Thanks to Daniel Bücken for assistance during preparation of the last set of XRD samples. Many thanks to Ivy for her assistance in the US and a very nice trip through many State- and National Parks.

The continuous improvements and changes in sample preparation for all purposes by Thomas Derichs are highly appreciated.

Pieter Berthier and Dennis Künkel are thanked for their time, discussions, and work preparing the XRD samples and results. Especially Pieter's input on the various types of clays in the active depositional environments and discussions on clay minerals in general is highly appreciated.

Additional thanks to the contributors to the iLoPS project 2, these are Alexandra Amann-Hildenbrand and Berhard Krooss from LEK at RWTH Aachen University and Norbert Klitzsch and Thomas Hiller from GGE at RWTH Aachen University.

The support during SEM analysis at the Central Facility for Electron Microscopy, GFE, RWTH Aachen University, by Alexander Schwedt and Fabian Mariano are highly appreciated. ESEM analyses were conducted at the Institute of Pathology at the University Hospital Aachen, RWTH Aachen University with the assistance of Stephan Rütten. Earlier electron microscopic analysis and BIB preparation was provided by Uwe Wollenberg, Geological Institute, RWTH Aachen University and Joyce Schmatz, Exogene Dynamik, RWTH Aachen University. Uwe's assistance during cathodoluminescence microscopy is highly appreciated as well. An additional thank you to Roman Klinghardt for thick section preparation for electron microscopy.

The discussions and breaks with Lars Gronen, IML, RWTH Aachen University, and the QEMSCAN analytics were a very good distraction from transmitted light microscopy to gain quantitative mineralogical data.

The operators of quarries are thanked for the granted permissions to acquire some samples, these are the Natursteinwerk Monser GmbH, Nordhorn, the Baustoffe Flechtingen, Bebertal, Alan of Cumbria Quarry Services, Penrith, U.K., and Mike of Eden Valley Stone Supplies, Armathwaite, U.K..

Finally, yet importantly I would like to thank my family and friends, whose time and support during the time in Aachen and transfer to Karlsruhe was invaluable, and who made it as pleasant as it was.

Abstract

Syntaxial overgrowth cementation, and thereby reservoir quality, can be affected by grain coating phases inhibiting nucleation. Reaction kinetics provide a means to model the development of cement phases over time. Additional algorithms constraining compaction behavior, porosity, and permeability development are used to model reservoir quality. Based on subsurface samples, these calibrated models can be used to model and predict reservoir properties in similar settings.

The additional incorporation of outcrop samples allows the study of spatial distribution of grain coatings, exceeding lateral dimensions covered by core material, and to delineate dependencies which can be applied for reservoir quality prediction modeling.

Silicate reaction kinetics provide a complimentary means to other established paleothermal indicators such as organic maturation for evaluating thermal reconstructions. Combined use of an organic maturation model with kinetic models for quartz and illite cementation were used to evaluate burial history scenarios for five sub-salt wells in lithologically and structurally complex Rotliegend reservoirs northeast of Hannover.

Models for organic maturation are most sensitive to maximum temperature and provide no direct evidence for the time of peak temperature or the overall duration of high temperatures. By contrast, the kinetics of quartz cementation is much more strongly influenced by the duration of exposure to high temperatures compared to organic indicators. Kinetic models for fibrous illite formation similarly are sensitive to time and temperature, and provide predictions for the time of illite formation that can be compared with K/Ar radiometric dates. Used collectively, these organic and inorganic paleothermal indicators provide improved constraints on thermal evolution compared to conventional approaches.

These indicators were used to evaluate two alternative burial history scenarios. Scenario one incorporates a hypothesized Jurassic heatflow peak together with significant Upper Jurassic deposition and subsequent erosion. Scenario two omits the Jurassic heat flow peak and the deposition and erosion of the Upper Jurassic. Although both of these scenarios are consistent with organic maturation data, scenario two leads to a far better match with quartz cement volumes and fibrous illite K/Ar dates.

The Penrith Sandstone Formation was chosen as an analog for other Rotliegend reservoirs. The half-Graben basin exposes the arid eolian depositional environments starting from alluvial conglomerates, via mixed fluvial-eolian deposits, to pure dune deposits. Differences of grain coating characteristics can be linked to grain size. In well-cemented sandstone samples, the grain coatings are better developed in finer grained laminae. Coarser grained laminae expose large amounts of syntaxial quartz overgrowth. The most prominent clay coating mineral is illite, which is often stained by hematite.

Structural features include normal faults and strike-slip faults of different ages. Diagenetic alterations around faults and deformation bands allow the delineation of relative temporal relations, revealing at least two different generations of deformation band formation associated to normal faulting. In the Vale of Eden succession one normal faulting event postdates burial diagenetic quartz cementation as evident by fault-focused fluid flow and associated leaching of iron and absence of quartz overgrowth.

Reservoir quality in deeply buried sandstones is strongly affected by grain coatings. To predict and infer the quality of undrilled reservoirs, constraints are needed to predict their occurrence. The objective of the study was to derive dependencies of the formation of grain coatings which are the basis for a universal predictive tool, i.e. applicable to the entirety of studied samples. Like other early diagenetic or even syn-depositional alterations, their presence and characteristics should be affected by the depositional environment and processes controlled thereby. The depositional environments, grain size, sorting, and the impact of volcanic rock fragments only locally correlate with the amount of grain coat coverage. On a regional to basin-scale (Southern Permian Basin and an adjacent sub-basin) variations emerge. Samples originate from northern Germany, the Netherlands, and the United Kingdom. This diminishes the availability of a universal predictive tool to assess reservoir quality based on sedimentary properties for the siliciclastic, fluvio-eolian Rotliegend in the Southern Permian Basin. The use of universal dependencies results in deviations during reservoir quality modeling and prediction. To prevent these deviations, it is imperative that the reservoir quality is estimated based on nearby analog wells in this Rotliegend setting.

Small-scale flow barriers in German Rotliegend siliciclastics were evaluated in regards to their appearance in reservoir intervals and their impact on fluid flow. Statistical evaluations and derivation of scaling relations lead to an improved understanding of these flow barriers in reservoir rocks. The link of petrophysical characteristics and petrographic observations results in a possible evaluation tool to assess the flow properties in reservoirs featuring deformation bands. Their formation, an interplay of chemical compaction and structural geology, can be inferred from petrographic interpretation.

Zusammenfassung

Syntaxiale Überwuchszementation, und somit Reservoirqualität, kann durch das Vorhandensein von Kornüberzügen, die eine Nukleation unterbinden können, beeinflusst werden. Anhand von Reaktionskinetiken der Zementation kann die Entwicklung der Zemente über die Zeit modelliert werden. Zusätzlich werden für die Modellierung der Reservoirqualitäten Algorithmen verwendet, die das Kompaktionsverhalten, sowie die Porositäts- und Permeabilitätsentwicklung beschreiben. Basierend auf Kernproben aus dem Untergrund können kalibrierte Modelle genutzt werden um Reservoirqualitäten in nahegelegenen Gebieten vorherzusagen. Die Korrelation mit Proben aus Aufschlüssen erlaubt die Analyse der räumlichen Verteilung von Kornüberzügen sowie die Herleitung von Abhängigkeiten, deren Kenntnis die Vorhersage der Reservoirqualitäten verbessert.

Die Kinetik der Quarzzementation kann ein zusätzliches Werkzeug zu anderen etablierten Paläotemperaturindikatoren, wie der Reifung organischen Materials, sein, die zur Evaluierung von thermischen Rekonstruktionen genutzt wird. In dieser Studie werden Modelle der organischen Reifung mit kinetischen Reaktionsmodellen der Quarz- und Illitzementation kombiniert, um Versenkungsrekonstruktionen für fünf sub-Salz Bohrungen in lithologisch und strukturell komplexen Rotliegend-Lagerstätten, nordöstlich von Hannover, zu evaluieren. Modelle zur Reifung organischen Materials reagieren am sensitivsten auf die maximal erreichte Temperatur und liefern keinen direkten Hinweis auf den Zeitpunkt des Erreichens oder die Residenzzeit im thermalen Milieu. Im Gegensatz dazu sind die Kinetiken der Quarzzementation deutlich sensitiver auf die Dauer, der Quarz hohen Temperaturen ausgesetzt ist. Die Bildung von fibrösem Illit ist ähnlich sensitiv auf Zeit und Temperatur. Zusätzlich können modellierte Phasen des Wachstums mit gemessenen K/Ar-Datierungen verglichen werden. In Kombination können diese organischen und anorganischen Paläotemperaturindikatoren die Rekonstruktion der thermischen Entwicklung im Vergleich zum konventionellen Ansatz verbessern.

Zwei Versenkungs- und Temperaturszenarien werden mit dieser Methode evaluiert. Das erste Szenario umfasst ein lokales Maximum des Wärmeflusses im Jura sowie die Ablagerung und anschließende Erosion der oberjurassischen Sedimente. Das zweite Szenario lässt beide Ereignisse außen vor. Obwohl die organische Reife beider Szenarien in Übereinstimmung modelliert wird, liefert Szenario zwei die bessere Übereinstimmung bei Quarzzementvolumina und K/Ar-Altern der Illite.

Die Penrith Sandstone Formation wurde als Analog für andere Rotliegendreservoire herangezogen. Im untersuchten Halb-Graben sind aride, kontinentale Ablagerungen von alluvialen Konglomeraten über gemischte fluvio-äolische Sandsteine bis hin zu reinen äolischen Dünen-Sandsteinen aufgeschlossen. Unterschiede der Kornüberzugscharakteristika können mit der Korngröße korreliert werden. In gut zementierten Sandsteinen sind die Tonkutane in feinkörnigen Laminen besser ausgebildet als in groben Lagen. Die häufigsten Tonkutane bestehen aus Illit, die durch das Auftreten von Hämatit rötlich gefärbt sind.

Das strukturgeologische Inventar besteht aus Abschiebungen und Blattverschiebungen unterschiedlichen Alters. Diagenetische Alterationen um Störungen und Deformationsbänder ermöglichen die relative Datierung unterschiedlicher Strukturen. Mindestens zwei diskrete Generationen von Deformationsbändern, die mit Abschiebungen assoziiert sind, können unterschieden werden. In der permischen Ablagerungssequenz des Vale of Eden entstand während der Versenkung der Einheiten bereits vor der Quarzzementation eine Deformationsband Generation. Eine zweite Generation folgte nach der Quarzzementation, welches durch die Abwesenheit von Quarzzementen und eine Bleichung der Tonminerale sowie aufgrund von Reduktion und Transport von Hämatit in Teilen der Formation, nachgewiesen wurde.

Reservoirqualität in tief versenkten Sandsteinen wird stark durch Kornüberzüge kontrolliert. Um deren Präsenz, und damit einhergehend die Reservoirqualität, vorhersagen zu können, wird ihr Auftreten analysiert. Während viele andere Aspekte, wie frühe Zementation, durch den Ablagerungsraum kontrolliert werden, können hier nur lokale Zusammenhänge zwischen Ablagerungsraum, Korngröße, Sortierung, und dem Einfluss von vulkanischen Lithofragmenten mit bestimmten Kutanabdeckungen festgestellt werden. Die Proben dieser Studie stammen aus vier Gebieten im Permischen Rotliegendbecken aus Deutschland, den Niederlanden und Großbritannien. Die hier bestimmten universellen Kutanabdeckungen resultieren in falschen Vorhersagen der Reservoirqualitäten. Eine genaue probenspezifische Analyse ist daher unersetzlich.

Kleinskalige Flussbarrieren in Rotliegend Sandsteinen in Deutschland wurden in Bezug auf ihr Vorkommen in Reservoirintervallen und ihren Einfluss auf Fließeigenschaften untersucht. Dabei erhöhen statistische Auswertungen und Ableitungen von Skalierungseffekten das Verständnis dieser Strukturen. Die Verknüpfung von petrophysikalischen Messungen und petrographischen Beobachtungen ermöglichen die Abschätzung eines möglichen Einflusses von Deformationsbändern auf das Fließverhalten. Basierend auf petrographischen Beobachtungen können dabei Kenntnisse über deren Entstehung gewonnen werden, bei welcher es sich um ein Zusammenspiel aus chemischer Kompaktion und Strukturgeologie handelt.

Contents

Declaration of Originality

Chapter 2: As first author, Benjamin Busch performed all data acquisition, diagenetic forward modeling, and re-evaluation of burial models. He wrote the entire article with inputs from Christoph Hilgers, Robert H. Lander, Linda Bonnell, and Dirk Adelmann. The discussions with all co-authors contributed to this manuscript.

Chapter 3: As first author, Benjamin Busch performed all sample analysis and evaluation. Assistance during QEMSCAN© analyses and evaluation by co-author Lars Gronen is highly appreciated. He wrote the article with inputs from Lars Gronen, Christoph Hilgers, and Dirk Adelmann. The discussions with all co-authors contributed to this manuscript.

Chapter 4: As first author, Benjamin Busch performed the sample analysis and evaluation. Analysis of a sub-set of sample was performed by Christina Schmidt and supervised and quality-controlled by the author. He wrote the entire article with inputs from Christina Schmidt, Christoph Hilgers, and Dirk Adelmann. The discussions with all co-authors contributed to this manuscript.

Chapter 5: As first author during his time as a PhD student, Benjamin Busch extended, combined, and re-evaluated the dataset provided to and prepared by him and Rebecca Winkler in their respective graduate projects. He wrote the entire article with input from Rebecca Winkler, Keyvan Osivandi, Georg Nover, Alexandra Amann-Hildenbrand, and Christoph Hilgers. The discussions with all co-authors contributed to this manuscript.

1 Introduction

1.1 Objective

The analysis of reservoir rocks (e.g. as reservoirs for water or hydrocarbons) has a focus on the present quality of the reservoir. Differences in detrital composition and authigenic phases impact the performance of reservoirs in terms of available pore space as storage space and permeability as means to produce the reservoir at a large scale. Flow anisotropies can be a result of both structural and diagenetic overprints at depth. All these processes can be combined into a reservoir quality analysis and modeling approach.

Reservoir quality analysis, modeling, and prediction has been a focus in recent years of both academia and the industry (Lander et al., 1997a; Lander et al., 1997b; Lander and Walderhaug, 1999; Walderhaug et al., 2000; Bloch et al., 2002; Makowitz et al., 2006; Lander et al., 2008; Ajdukiewicz and Lander, 2010; Ajdukiewicz et al., 2010; Lander and Bonnell, 2010; Taylor et al., 2010). The factors controlling the porosity and permeability of a given reservoir need to be studied in more detail to accurately predict properties based on a forward model. Most prominent impacts on reservoir productivity are cementation, compaction, and faulting. The first two are a function of burial depth and temperature, as well as available fluids carrying ions in solution. Faulting is a result of the tectonic setting and mechanical properties of the lithology. Constraining the impact of each of these processes on a series of samples from a certain reservoir or group of wells or reservoirs based on petrographic analysis, is the first step to a reservoir quality model using Touchstone™ diagenetic modeling software. During petrographic analysis, special attention is given to grain coating phases, since these are described to impact syntaxial overgrowth cementation by quartz (Heald and Larese, 1974; Wilson, 1992; Ehrenberg, 1993; Ryan and Reynolds, 1996; Aagaard et al., 2000; Storvoll et al., 2002; Billault et al., 2003; Ajdukiewicz and Lander, 2010; Ajdukiewicz et al., 2010; Mahaney et al., 2010; Taylor et al., 2010; Ajdukiewicz and Larese, 2012; Dowey et al., 2012; Haile et al., 2015). Where grain coatings are continuously present, quartz overgrowth is inhibited (Fig. 1.1 a). However, if the grain coating is incomplete (broken) or absent, syntaxial quartz overgrowth can occur (Fig. 1.1 a, b). The incorporation of granulometric data, petrophysical and geochemical analyses are other important informations to increase the quality of the model.

Cementation kinetics of quartz and illite are closely linked to burial depth and temperature. Opposed to this, compaction is modeled based on different rigidities of detrital and authigenic phases and their response to burial depth. During modeling, alterations in the mineral content during diagenesis will ultimately impact the results of the model. New mineral phases alter permeability, porosity, and available intergranular pore space during cementation. These new minerals will also impact the compaction behavior, which will be taken into account during modeling. The result is a model, which, based on burial and thermal reconstructions, can reproduce relevant data for reservoir quality analysis. Upon completion of the diagenetic model, the paragenetic sequence and compaction behavior of the grains can be used to model and predict reservoir quality away from well control in e.g. nearby undrilled areas, given a good estimate on burial conditions and lithologies.

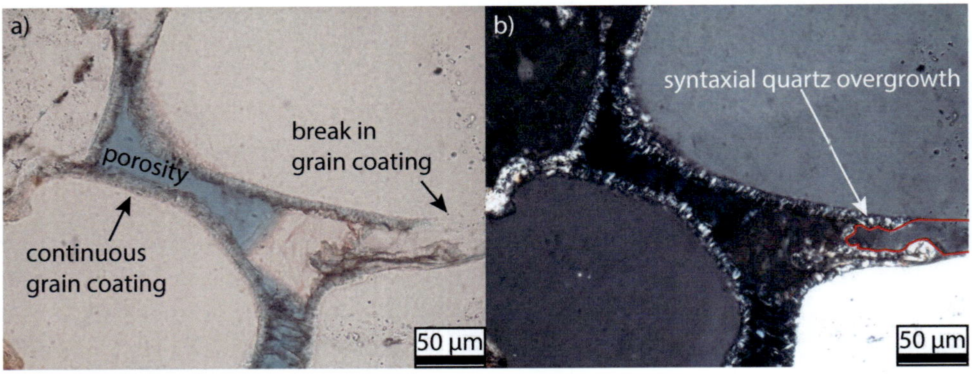

Figure 1.1: Schematic impact of grain coatings in north German reservoir rocks from the Rotliegend. a) Photomicrograph in plane polarized light, where complete and incomplete grain coatings are shown. b) Photomicrograph in cross polarized light where syntaxial quartz overgrowth (red outline) can be observed to have nucleated at a break in a grain coating.

1.2 Aims

The aim of this study is to better understand the distribution and effect of grain coatings on reservoir properties in the Rotliegend. Therefore, their effect on syntaxial quartz overgrowth cementation is combined with a diagenetic forward modeling approach. Additionally, structural diagenetic alterations are incorporated. The focus during this work is on four major aspects:

- Showing the differing sensitivities of reaction kinetics to thermal exposure, embedded in the diagenetic modeling software, to add another calibration perspective to both basin- and diagenetic modeling.
- Displaying the benefit of incorporating grain coatings in the evaluation of structural- and diagenetic alterations in sandstones.
- Linking predictable or reconstructable properties and controlling parameters on the presence and extent of grain coatings.

- Coupling the evolution of microstructures to sedimentary texture, diagenetic alterations, and petrophysical alterations.

1.3 Overview of the thesis

1.3.1 Organic and inorganic reaction kinetics (Chapter 2)

The kinetic calculation of quartz and illite cement phases is highly sensitive to changes in thermal exposure and rely on accurate thermal reconstructions for the studied areas. Their sensitivity to thermal exposure at e.g. a constant temperature differs largely from the sensitivity of organic matter maturity to constant temperature exposure (Lander et al., 2008; Lander and Bonnell, 2010; Le Bayon et al., 2011). Both organic (vitrinite reaction kinetics) and inorganic (describing quartz and illite kinetics) algorithms on their own have been proven to accurately reproduce experimental and natural sample series (Lander et al., 2008; Lander and Bonnell, 2010; Le Bayon et al., 2011; Le Bayon et al., 2012; Harwood et al., 2013). However, a large mismatch during diagenetic modeling might result, should the thermal reconstruction be inaccurate. In chapter 2, we show the added value in using both organic maturation and inorganic cementation (quartz and illite) algorithms to find the thermal reconstruction which can satisfy all of the three criteria. This combined approach will ultimately improve the assessment of settings where thermal indicators are rare or results are complex. The database and samples are from five wells penetrating the reservoir interval of the Bahnsen Sandstone Member of the Permian Rotliegend northeast of Hannover, Germany. Samples and data were provided by Wintershall Holding GmbH.

1.3.2 Grain coatings and structural diagenesis (Chapter 3)

The occurrence of grain coating phases is a factor that will ultimately impact the petrophysical characteristics and diagenetic alterations that can be observed in a reservoir. Where grain coatings are present and cover a large percentage of grain surfaces, syntaxial overgrowth cementation is inhibited. The larger the grain coated area, the smaller the available surface area for quartz nucleation on substrate grains. Where grain coats are present but broken or only cover a small amount of the grain surfaces, syntaxial overgrowth cementation is only inhibited locally, whereas some parts of the grains feature extensive overgrowth cementation. In these cases, the grain coats are readily overgrown. To capture the heterogeneity in the distribution of clay coatings in fluvio-eolian settings, an analog study was prepared in the Permian Rotliegend of northern England. Fluvio-eolian deposits from the Penrith Sandstone and the Brockrams of the Appleby Group have been studied in the Vale of Eden half-Graben in the area from Brough in the South to Lazonby in the North.

In chapter 3, we analyze and discuss the impact of clay coatings and their distribution in the lithology, the application of state-of-the-art electron microscopy analyses, and the possible additional impact of clay coatings in this setting to relatively date deformation and possible compartmentalization by deformation bands and their interaction with diagenesis as a response to different fluid chemistries. In all these aspects, grain coatings are essential.

1.3.3 Distribution of grain coatings (Chapter 4)

To deduce the heterogeneity and to possibly delineate controlling factors on the distribution of grain coatings in fluvio-eolian sandstones of the Rotliegend in central and northwestern Europe, samples from multiple areas across the Southern Permian Basin were compiled. Samples from the Bahnsen Sandstone Member (5 wells northeast of Hannover), the Slochteren Sandstone Formation (5 wells in the K10 and K13 blocks offshore the Netherlands), the Penrith Sandstone Formation (outcrops in the Vale of Eden, Cumbria, United Kingdom), and the Parchim Formation (outcrops near Bebertal, Saxony-Anhalt, Germany) have been analyzed. Core material and samples from NE-Hannover were provided by Wintershall Holding GmbH. Samples and core material from the Dutch offshore were provided by Wintershall Noordzee B.V. In chapter 4, we analyze and discuss the impact of the depositional environment, granulometry, detrital composition, and the diagenetic overprint on the presence and extent of grain coatings.

1.3.4 Deformation bands in the Rotliegend (Chapter 5)

Deformation bands in siliciclastics from the Rotliegend of Northern Germany are discussed to present flow barriers in hydrocarbon reservoirs. Their microstructure poses a heterogeneity as opposed to the undisturbed host rock. In chapter 5, we evaluate petrographic observations, which can deduce a possible evolution of cataclastic deformation bands in these sandstones. Statistical evaluations based on core material and image logs (Formation Micro Images, FMI) based on resistivity measurements are a means to infer and validate larger scale tectonic features and scaling relations. Their petrophysical characterization delineates a systematic impact on localized flow properties in reservoir rocks. Samples were provided by RWE Dea AG (now DEA Group).

1.4 Parts of this thesis which have been published

Articles:

Busch, B., Winkler, R., Osivandi, K., Nover, G., Amann-Hildenbrand, A., and Hilgers, C., 2015, Evolution of small-scale flow barriers in German Rotliegend siliciclastics, *in* Armitage, P. J., Butcher, A. R., Churchill, J. M., Csoma, A. E., Hollis, C., Lander, R. H., Omma, J. E., and Worden, R. H., eds., Reservoir Quality of Clastic and Carbonate Rocks: Analysis, Modelling and Prediction, Volume 435: London, Geological Society, Special Publication.

Accepted manuscripts:

Busch, B., Hilgers, C., Lander, R. H., Bonnell, L. M., and Adelmann, D., Reservoir quality and burial model evaluation by kinetic quartz and illite cementation modeling: Case study Rotliegend, N-Germany, accepted for publication in AAPG Bulletin.

Submitted manuscripts:

Busch, B., Hilgers, C., Gronen, L., and Adelmann, D., subm., Cementation and structural diagenesis of fluvio-eolian Rotliegend sandstones, North England.

Busch, B., Hilgers, C., Schmidt, C., and Adelmann, D., in prep., Distribution of clay coatings in Permian fluvio-eolian red beds, case study Rotliegend, Southern Permian Basin.

Conference contributions:

Busch, B., Winkler, R., Osivandi, K., Nover, G., and Hilgers, C., Integrated interpretation of small-scale flow barriers on resistivity logs to validate and improve seismic-scale faults - first results, at DGMK Frühjahrstagung, Celle, 2014.

Busch, B., Winkler, R., Osivandi, K., Nover, G., Amann-Hildenbrand, A., and Hilgers, C., Evolution of small-scale flow barriers in German Rotliegend siliciclastics, at Reservoir Quality of Clastic and Carbonate Rocks: Analysis, Modelling and Prediction, Geological Society, London, 2014.

Busch, B., Hilgers, C., Adelmann, D., Lander, R. H., and Bonnell, L. M., Impact of Clay Coats on Reservoir Heterogeneity and Reservoir Quality in Permian Red Beds, at DGMK Frühjahrstagung, Celle, 2015.

Busch, B., Reservoir-Heterogenitäten und deren Modellierung zur RQ-Vorhersage, at Aachener Geotag – Energie-Reservoire Untertage: Exploration, Speicherung, Ökonomie, Aachen, 2015.

Bamberg, B., Kaiser, L. Busch, B., Hilgers, C., Analog study of Rotliegend reservoir rocks in the Vale of Eden, UK, at Aachener Geotag – Energie-Reservoire Untertage: Exploration, Speicherung, Ökonomie, Aachen, 2015.

Busch, B., Schmidt, C., Hilgers, C., and Adelmann, D., 2016, Distribution of coatings in fluvio-aeolian red beds and their impact on reservoir quality modeling, at EAGE Conference and Exhibition: Vienna, 2016.

Schmidt, C., Hilgers, C., Busch, B., Petrophysical and diagenetic heterogeneities in siliciclastic reservoir rocks, at GeoTirol 2016, Annual Meeting DGGV, Innsbruck, 2016.

Busch, B., Hilgers C., Lander, R.H., Bonnell, L.M., D. Adelmann, C. Schmidt, Impact of grain coats on quartz cementation, reservoir properties, and modeling, at Relevance of Clastic and Carbonate Diagenesis for Reservoir Quality, Force, Stavanger, 2016

2 Reservoir quality and burial model evaluation by kinetic quartz and illite cementation modeling: Case study Rotliegend, N-Germany

2.1 Abstract

Silicate reaction kinetics provide a complimentary means to other established paleothermal indicators such as organic maturation for evaluating thermal reconstructions. In this study we combine the use of an organic maturation model with kinetic models for quartz and illite cementation to evaluate burial history scenarios for five sub-salt wells in lithologically and structurally complex Rotliegend reservoirs.

Models for organic maturation are most sensitive to maximum temperature and provide no direct evidence for the time of peak temperature or the overall duration of high temperatures. By contrast, the kinetics of quartz cementation are much more strongly influenced by the duration of exposure to high temperatures compared to organic indicators. Kinetic models for fibrous illite formation similarly are sensitive to time and temperature and provide predictions for the time of illite formation that can be compared with K/Ar radiometric dates. Used collectively these organic and inorganic paleothermal indicators provide improved constraints on thermal evolution compared to conventional approaches.

In this study we use these indicators to evaluate two alternative burial history scenarios. Scenario one incorporates a hypothesized Jurassic heatflow peak together with significant Upper Jurassic deposition and subsequent erosion. Scenario two omits the Jurassic heat flow peak and omits the deposition and erosion of the Upper Jurassic. Although both of these scenarios are consistent with organic maturation data, scenario two leads to a far better match with quartz cement volumes and fibrous illite K/Ar dates.

2.2 Introduction

Diagenetic studies are used to establish reservoir quality from core material (Schöner and Gaupp, 2005; Gier et al., 2008). While such studies are essential for understanding the controls on diagenesis and reservoir quality they are not in a strict sense predictive. Consequently, forward modeling of diagenetic processes, in concert with rock characterization, has become a focus in hydrocarbon exploration and production as a means to make quantitative reservoir quality predictions for undrilled locations (Lander and Walderhaug, 1999; Makowitz et al., 2006; Lander et al., 2008; Ajdukiewicz and Lander, 2010; Taylor et al., 2010). Such reservoir quality prediction modeling relies on the available knowledge of physical and chemical properties of rocks, kinetics of mineral growth, and reconstructed burial histories.

The burial history reconstruction is usually based on information on lithologies including thickness, distribution, and erosion maps as well as on the maturity of organic material (i.e. vitrinite reflectance) and has successfully been applied to many fields (Littke et al., 2008; Uffmann and Littke, 2011). However, as shown experimentally by Le Bayon et al. (2012), when exposed to high temperatures and pressures vitrinite reflectance increases within minutes to hours while increasingly higher pressures retard organic maturation. Longer exposure to high temperatures and pressures result in only slight changes of the attained vitrinite reflectance (Le Bayon et al., 2012). Quartz cement growth on the other hand is shown to follow kinetic relations that are much more sensitive to the duration exposed to high temperatures (Walderhaug, 1994, 2000; Lander et al., 2008). If exposed to a constant temperature over a long period of time the formation of quartz cement volumes increases nearly linearly with time (Makowitz et al., 2006) as opposed to a significant initial increase and then a decrease in reflectance development as observed in vitrinite (for T (temperature at the end of heat up) > ~270 °C; Le Bayon et al., 2012). Thus, short high temperature pulses such as those associated with transient fluid-flow events (e.g. Sindern et al., 2007) tend to have a larger influence on vitrinite reflectance (provided the temperature is the maximum historical value) compared to quartz cement volume.

The rate of quartz overgrowth cementation is primarily controlled by the area available for overgrowth nucleation, temperature, and the precipitation kinetics. The nucleation surface area for quartz overgrowths is a function of the quartz grain abundance and size distribution and generally diminishes in response to the occurrence of grain coatings (Heald and Larese, 1974; Lander and Walderhaug, 1999; Walderhaug, 2000; Ajdukiewicz and Larese, 2012; French et al., 2012). Nucleation surface area also decreases as contact areas between grains increase in association with compaction and as quartz and other cement types fill inter granular pores (Lander et al., 2008).

Two approaches have been used for deriving parameters describing the precipitation kinetics: (1) laboratory experiments (e.g. Rimstidt and Barnes, 1980) and (2) analysis of geological constraints (Walderhaug, 1994, 1996; Lander and Walderhaug, 1999; Walderhaug, 2000; Walderhaug et al., 2000; Lander et al., 2008). Of these approaches the latter has proven to be a far more accurate basis for pre-drill prediction of quartz cement abundances (Ajdukiewicz and Lander, 2010). An additional factor of significance for the kinetics of quartz growth is the reduction in growth rate that occurs with the progressive development of euhedral crystal faces (Lander et al., 2008). This euhedral effect results in reduction of the surface area normalized rate with diminishing crystal domain size. Thus the rate tends to slow as grain size becomes finer for monocrystalline grains and to be lower for polycrystalline grains compared to monocrystalline grains of comparable size (Waugh, 1970a; Lander et al., 2008).

Although previous studies have considered the application of quartz cement abundance as a thermal indicator (Awwiller and Summa, 1997; Lander et al., 1997a; Lander et al., 1997b; Perez et al., 1999; Matthews et al., 2002) in this study we employ an additional inorganic thermal indicator associated with the reaction of kaolinite and a potassium source (assumed to be K-feldspar grains in this study) to form fibrous illite in sandstones (Lander and Bonnell, 2010). The fibrous illitization kinetic model, like the quartz cementation kinetic model, is sensitive to both time and temperature. A difference between these models, however, is that whereas quartz cementation appears to be a continuous process so long as nucleation area is available, the illitization reaction stops once the kaolinite or potassium source is consumed. Additionally, this model has the unique ability to provide predictions for the timing of illite formation that may be compared to K/Ar radiometric dates.

In this study we use diagenetic models to evaluate burial and thermal exposure models of a Rotliegend reservoir. The burial and thermal models are used to define the pressure and temperature conditions controlling the diagenetic development. By using a well calibrated diagenetic model based on mineral kinetics and physical and chemical properties of the rocks, we test if a given thermal and burial model can reproduce the measured values in quartz cement volumes and illite ages. Additionally, we model the intergranular volume (IGV), porosity, and permeability evolution over time and compare these with core data. These models may improve burial reconstructions with an impact on organic matter maturation scenarios as well as provide insights into reservoir quality controls, improving play analysis.

2.3 Geological Setting

Samples are taken from five wells penetrating Rotliegend strata (Bahnsen Sandstone Member, Fig. 2.1) in the North German Basin north of Hannover (Fig. 2.2). Sediments were deposited in fluvio-eolian settings within a Permian graben structure and display a wide range of porosities, permeabilities, and cement types. The sampled depositional environments include dunes (n = 11); dry- (n = 34), damp sandflats (n = 4); and one lacustrine sample (n = 1). The Rotliegend is overlain by Zechstein salt layers and one of the five wells penetrates the Rotliegend strata underneath a salt pillow (well C). All five wells are within a radius of 5 km from well C (Fig. 2.2). The reservoirs are currently at depths ranging from 4264 to 4570 m TVD.

Authigenic illite samples dated from Rotliegend strata from the North German Basin using the K/Ar method yield ages of 190 to 155 Ma (Zwingmann et al., 1998) and ~190 Ma (Platt, 1993).

Group		Subgroup	Formation	Member	Age
Rotliegend	Upper Rotliegend	Elbe	Hannover	Heidberg	258
				Munster	
				Niendorf	
				Dambeck	
				Bahnsen	
				Wustrow	
				Ebstorf	260
			Dethlingen	Einloh	
				Strackholt	
				Schmarbeck	
				Wettenbostel	
				Garlstorf	
				Findorf	
				Deth. Sands	262
		Havel	Mirow		264
			Parchim		266
		Müritz			
	Lower Rotliegend				296

Figure 2.1: Stratigraphic subdivision of the Rotliegend in N-Germany (after Schöner (2006)), stratigraphic ages from Menning (1995)). The Bahnsen Sandstone Member which is the focus of this study is part of the Hannover Formation.

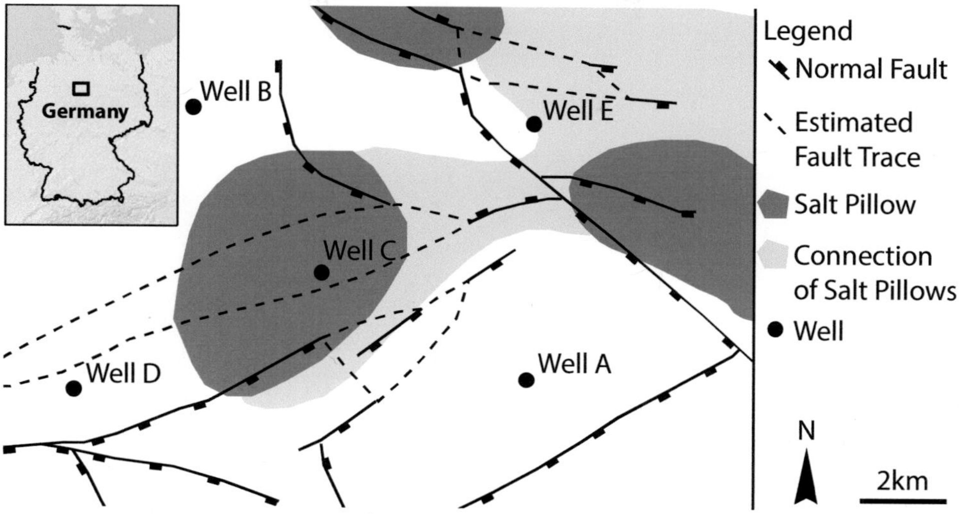

Figure 2.2: Location of the five studied wells including the structural inventory and location of salt pillows and their connection. The dashed lines are inferred normal faults.

2.3.1 Burial models

The only paleothermal constraint used during the initial development of the provided 1D scenario 1 thermal reconstruction is vitrinite reflectance data (Fig. 2.3 a, b) (Peisker, 2013). This scenario produces very high temperatures during the Jurassic, reaching maximum temperatures of ~220 °C in the Rotliegend strata of well A and 196 °C in well B, as well as a second period of high temperatures during the Paleocene (Fig. 2.3 b).

This scenario considers heat flow pulses associated with four extensional phases during the (1) Lower Triassic, (2) Upper Triassic, (3) Middle Jurassic, and (4) Upper Jurassic as based on general reconstructions for the Southern Permian Basin (Fig. 2.4, heat flow scenario 1) (Pharaoh et al., 2010). The heat flow values for this reconstruction are based on general values from Allen and Allen (2005). The high heat flow values at the beginning of the Permian are associated with basin formation and early Permian volcanism (Peisker, 2013). The heat flow increase during the Jurassic (~190 Ma) is associated with deposition following North Sea rifting and is interpreted to be additionally controlled by North Sea Doming (cf. Pharaoh et al., 2010). In between rifting events the heat flow decreases to approximately 60 $\frac{mW}{m^2}$ to attain typical continental heat flow values (Peisker, 2013).The decline in heat flow from the Eocene onwards is defined to match present-day temperatures and heat flow of approximately 55 $\frac{mW}{m^2}$ (Peisker, 2013).

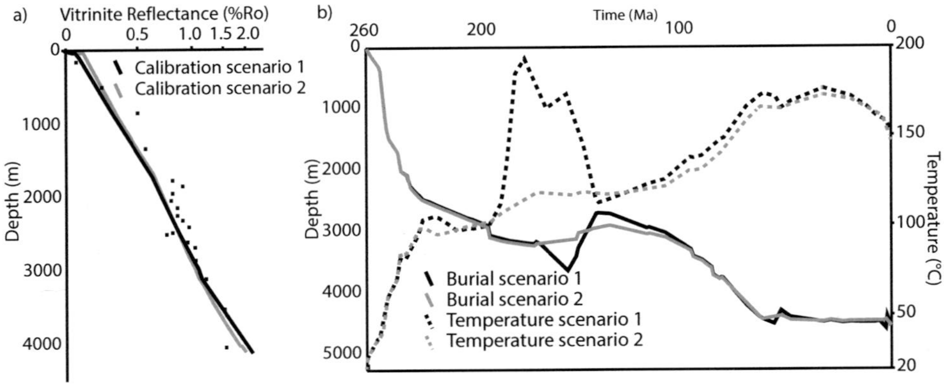

Figure 2.3: (a) Vitrinite reflectance modeled with EASY%R_0 for two subsidence scenarios in PetroMod™ 2012. The overall performance of modeled vitrinite reflectance is comparable for the two scenarios. (b) The comparison of two burial scenarios for well B highlight the small difference in burial reconstructions (solid lines). The deposition and erosion of Upper Jurassic strata is omitted in scenario 2 as opposed to scenario 1. The maximum temperatures (dashed lines) attained during burial differ strongly reflecting significant differences in heat flow.

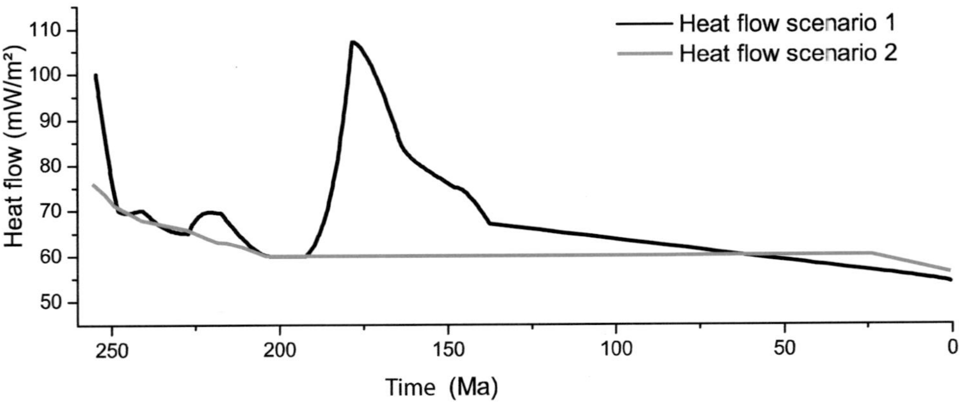

Figure 2.4: Comparison of heat flow scenarios for well B. Heat flow scenario 1 incorporates a peak between 200 and 145 Ma associated with Jurassic rifting and North Sea doming and is responsible for the modeled temperature peak in scenario 1 (Peisker, 2013). This scenario also incorporates a period of Upper Jurassic deposition and subsequent erosion (burial scenario 1, Fig. 2.3 b). Heat flow scenario 2 from Uffmann and Littke (2011) assumes that the Jurassic stretching and doming event did not affect the study region in terms of heat flow changes. Likewise, the associated burial scenario 2 adjusts Lower and Middle Jurassic deposition, as well as erosion during the Upper Jurassic (Fig 2.3b.).

Based on vitrinite reflectance development assumptions, 700 m of Upper Jurassic deposition and erosion in the study area on the Pompeckj Block was added to account for the measured vitrinite reflectance data (Peisker, 2013). The amount is based on correlations of the deposition within the adjacent Lower Saxony Basin (Peisker, 2013), which was not affected by Late Jurassic Uplift. This eroded thickness is added to the deposition and erosion of approximately 650 m of Lower and Middle Jurassic strata in a nearby region (cf. Schwarzer and Littke, 2007). Deposition resumed in the study area during a transgression in the Albian.

The erosion during the Upper Cretaceous and early Tertiary inversion is represented by the incomplete development of Upper Cretaceous strata. Thus, a comparative set of 1D burial models (scenario 2) was prepared for this study (Fig. 2.3, 2.4).

The lower temperature scenario 2 features a heat flow scenario derived from 1D, 2D, and 3D burial models published in Schwarzer and Littke (2007) and Uffmann and Littke (2011).

The heat flow reconstruction is characterized by a decrease in heat flow following Permian rifting and volcanism to attain a typical continental heat flow around 60 $\frac{mW}{m^2}$, which is further reduced to 55 $\frac{mW}{m^2}$ to match present day values and temperatures on the Pompeckj Block (Fig. 2.4) (Schwarzer and Littke, 2007; Uffmann and Littke, 2011; Bruns et al., 2013). The heat flow reconstructions in the region following Permian rifting feature a decrease to a range of typical continental heat flow in the Mid-Jurassic, assuming gradual cooling from theoretical stretching models (McKenzie, 1978; Schwarzer and Littke, 2007). The effect of North Sea doming and rifting on the heat flow was neglected due to the distance to the studied area and absence in other models of the region (Schwarzer and Littke, 2007; Uffmann and Littke, 2011). The temperature increase is thus not as large as in scenario 1. The maximum temperature reached in this scenario is 172 °C and was attained during the phase of deepest burial during the Paleocene. By contrast, in scenario 1 the maximum temperature occurred during the Jurassic. The vitrinite reflectance and present day temperature results from scenario 2 provide a comparable match to measured values to that of scenario 1 (Fig. 2.3 a, b). Importantly, the higher temperatures in scenario 1 resulting from the heat flow peak during the Jurassic are not necessary to match the maturity of organic matter (Fig 2.3 a). Scenario 2 does not accommodate for the amount of deposition and subsequent erosion during the Upper Jurassic as assumed in scenario 1 and assumes an extended hiatus until the Albian transgression where Lower and Middle Jurassic strata are still present. A localized erosive event during the Upper Jurassic as presented in Schwarzer and Littke (2007) was added to match the presence or absence of Lower and Middle Jurassic strata in each well (cf., Baldschuhn et al., 2001; Bombien et al., 2012).

2.4 Methods

Modal compositions were established from point counting (300 counts in total) petrographic thin sections and are used to constrain the amount of reactants and products of diagenetic reactions. The measurement of grain coat coverages of quartz grains was performed on 50 grains per sample. In total 50 thin sections and associated plug samples were used as a basis for this study.

The modeling suite Touchstone™, developed by Geocosm LLC, encompasses algorithms to model the kinetics of quartz cement and fibrous illite growth (Lander et al., 2008; Lander and Bonnell, 2010) as well as compaction and the occurrence of various other cement types. Analog data from core material is used for this diagenetic forward modeling approach for reservoir quality modeling and prediction in sandstones. The kinetic parameters for each reaction are based on geologic constraints rather than experimental data. In the case of quartz cementation the primary kinetic parameter of interest is an activation energy (E_A) for crystal growth including changes in crystal growth rate depending on grain size and crystal domain size (Lander et al., 2008). The adjustable kinetic parameter for the fibrous illite model is an activation energy for the reaction of kaolinite and a potassium source (K-feldspar in this study) to form illite (Lander and Bonnell, 2010).

Additional model parameters constrain physical properties of detrital and authigenic phases related to their compaction behavior, as well as porosity and permeability. Compaction and the intergranular volume are modeled by using a proprietary algorithm (Makowitz et al., 2006; Lander et al., 2008) that builds on the concepts of Lander and Walderhaug (1999). The timing of the formation of authigenic phases in the paragenetic sequence can be constrained by thermal-, temporal-, or burial-parameters (depth) and impacts the compaction behavior and the development of the intergranular volume (IGV) over time. The IGV is typically measured by point counting from petrographic thin sections and is the sum of intergranular porosity, cements and depositional matrix (Houseknecht, 1987; Paxton et al., 2002). The IGV calibration is based on the impact of rigidity of authigenic and detrital mineral phases and their contribution to the IGV development (Lander and Walderhaug, 1999).

Additionally, the kinetic model of the alteration of kaolinite and feldspars to illite (Lander and Bonnell, 2010) has been used for simulations with Touchstone™. Measured K/Ar-ages of authigenic illite are used to evaluate the modeled time of formation as given by the diagenetic model. A deviation of the modeled illite ages from measured data would require an improved thermal reconstruction of the subsidence model or an adjustment of diagenetic model parameters.

Parameter optimization simulations have been carried out with respect to the activation energies described above as well as for compaction characteristics for the rock's solid components. This process is performed by iteratively changing parameters and using a fitness function of the mean square difference between measured and simulated quartz cement volumes, illite cement volumes and the IGV (Lander et al., 2008). Outputs of an optimization simulation are a parameter suite that lead to an optimal match with measured values.

We used the quality of the optimization results as a means to evaluate the geological veracity of the two heat flow scenarios (Fig. 2.3, 2.4). The heat flow scenarios were used to create two sets of 1D burial models for each of the five wells. The paleosurface temperatures in both scenarios are based on reconstructions by Wygrala (1989) and are calculated by PetroMod™ depending on the present day latitude. The lithotypes as defined by Peisker (2013) were not modified. The paleobathymetry is derived from facies maps from Ziegler (1990) and estimated from well logs (Peisker, 2013). The performance of both burial models with the same kinetic parameters for vitrinite maturation (EASY%R_0, Burnham and Sweeney (1989)) is comparable (Fig. 2.3), although the thermal exposure in scenario 1 is much higher (Fig. 2.4). The thermal exposure encompasses the time exposed to temperatures above which significant quartz cementation occurs.

2.5 Results

2.5.1 Petrography

Rock samples (n = 50) contain between 48 and 69 % detrital quartz grains including polycrystalline quartz grains, 4 to 15 % feldspar grains (including plagioclase), and up to ~8 % rock fragments (Tab. A.1, Fig. 2.5 a-f). Authigenic phases varying in content in different samples comprise ~2 to 23 % quartz cement, up to 3 % K-feldspar cement, up to 14 % illite, up to 9 % chlorite, up to 16 % calcite cements (Fig. 2.5 b, d), and up to 20 % sulfate cements including anhydrite (Fig. 2.5 f), gypsum and barite (Tab. A.1). Grain coating materials only present a small volumetric fraction of the samples (mean of 0.3 %) and the main coating phases are chlorite and smectite-chlorite mixtures. The surface area of grains coated by either clay or hematite ranges from 5 to 97 %. Samples with high grain coat coverages show substantially lower quartz cement abundances (Fig. 2.5 a, b), compared to samples with poorly developed coatings (Fig. 2.5 c, d, e). There is no apparent correlation between depositional environments and detrital composition, authigenic minerals, or grain coating coverage. The IGV of the sample set ranges from 16 to 39 %. Higher values are associated with early carbonate and sulfate cements. The petrophysically derived porosity (Φ) ranges from 0.9 to 20.6 % and permeability (k) ranges from 0.01 to 122 mD. A point count compilation for each sample, including petrophysical parameters can be found in the appendix (Tab. A.1).

Figure 2.5: Photomicrographs of some representative Rotliegend samples of dry sandflat deposits from the studied wells illustrate the different reservoir cementation types. (a) and (b) Sub-salt sample containing quartz and feldspar grains in plane- (ppl) and cross polarized light (xpl), respectively. Significant intergranular porosity is retained by the presence of thin, faint greenish chlorite-smectite coatings, showing white birefringence (Well C). The coating phase, however, does not inhibit the growth of carbonate cement phases (arrow, top left). (c) (ppl image) and (d) (xpl image) Hematite dust rims and only minor amounts of illitic coatings did not inhibit the growth of syntaxial quartz cements (Well E). (e) Pervasive quartz cementation occurs in the absence of grain coating phases (ppl image) Well B). (f) Authigenic cement phases such as anhydrite and carbonates occlude parts of the pore space in samples from every well (xpl image, Well C).

2.5.2 Quartz kinetics calibration

Thermal histories from the two heat flow models are used to compare the amount of quartz and illite cements established from core material with modeled quartz and illite cement volumes. The quartz cementation kinetic parameters include an activation energy (E_A) for crystal growth along the c-axis on non-euhedral surfaces and a pre-exponential parameter (A_O) (Lander et al., 2008). This growth model is based on the Prism2D algorithm for surface area normalized crystal growth (Lander et al., 2008; Lander and Laubach, 2014) During parameter optimization we allowed the E_A value to vary between 55 to 65 $\frac{kJ}{mol}$ and fixed the A_O value at $9*10^{-12}\ \frac{mol}{\frac{cm^2}{s}}$. These values are consistent with ranges reported by Lander et al. (2008) for quartz cementation in sandstones from four geologically diverse datasets. All samples within the given scenario are subject to the same kinetic parameters.

We obtained an optimized E_A value of 58.2 $\frac{kJ}{mol}$ in the higher-temperature heat flow model (scenario 1). The calculated quartz cement values deviate from measured values by a maximum of 12 vol%. The quartz cement values tend to be overpredicted for wells B, D and E but underpredicted for sub-salt well C (Fig. 2.6 a). The only good fit for this model scenario is for samples from well A. To create a reasonable fit of measured vs. calculated data for the other wells, we conducted parameter optimization for each well individually, disregarding samples from other wells. The quartz activation energies obtained by this method ranged from values exceeding 65 $\frac{kJ}{mol}$ for wells B, D, to less than 55 $\frac{kJ}{mol}$ for well C. We deem this thermal scenario to be inconsistent with the quartz cement abundances in these wells due to the combination of (1) poor model performance for the case where the same kinetic parameters are used for all wells, (2) optimized E_A values derived from individual wells that are widely divergent despite close proximities, and (3) E_A values from individual wells that are outside of the ranges reported elsewhere (i.e., Makowitz et al., 2006; Lander et al., 2008).

By contrast, the lower-temperature scenario 2 results in modeled quartz cement values that are in close correspondence with measured values from thin section analysis (Fig. 2.6 b). The derived E_A for quartz cementation of 58.2 $\frac{kJ}{mol}$ is used for samples from all wells and is well within the range of values reported from other datasets. The quartz cement volumes are well reproduced in samples from all represented depositional environments.

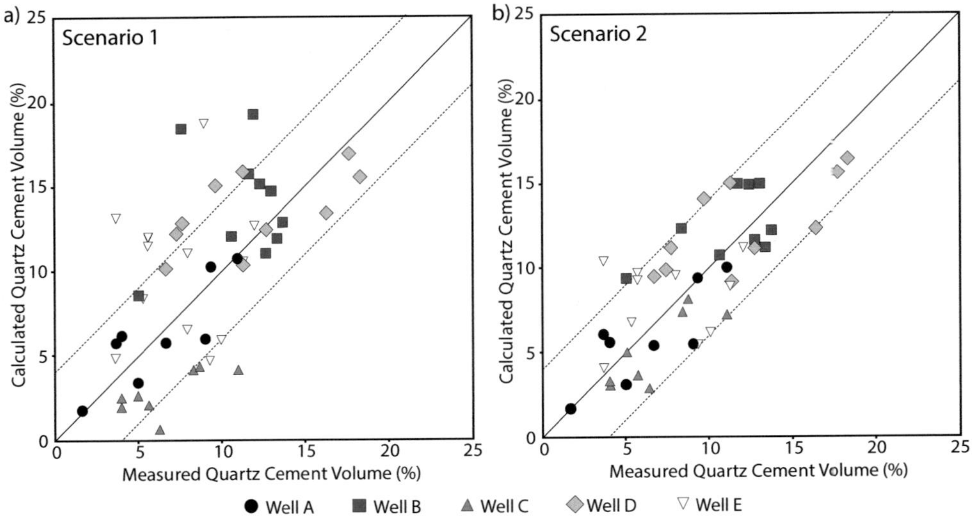

Figure 2.6: Comparison of the performance of the quartz cement model for the two thermal history scenarios. (a) The high temperature scenario 1 results in a systematic overprediction of generated quartz cements for wells B, D and E. The deviation from measured amounts of quartz cement is up to 12 %, well outside of the likely measurement uncertainty of around four percent (dotted lines). However, the subsalt dataset of well C underpredicts the amount of measured quartz cement by up to 7 %. (b) Modeled quartz cement values match data the sample set within the likely measurement uncertainty for most samples of ±4 %.

2.5.3 Illite kinetics calibration

As an additional evaluation of the two alternative thermal scenarios we also evaluate the performance of a kinetic model for illite formation. In this case we use the kinetic constraints from Lander and Bonnell (2010).

For the high temperature scenario 1 the formation of illite is predicted to be in a range between 245 and 195 Ma (Fig. 2.7 a), which partly predates the measured values (Fig. 2.7 b) of Zwingmann et al. (1998). By contrast, the start of illite formation in scenario 2 (Fig. 2.7 c) is modeled to be at 210 Ma and to have ceased at 155 Ma, which is consistent with measured K/Ar-ages of illite (Platt, 1993; Zwingmann et al., 1998, 1999) in the North German Basin on the Pompeckj Block (Fig. 2.7 b). Peak formation rates match the measured K-Ar ages of illite in a nearby Rotliegend graben of 190-155 Ma (Fig. 7 b) (Zwingmann et al., 1998)

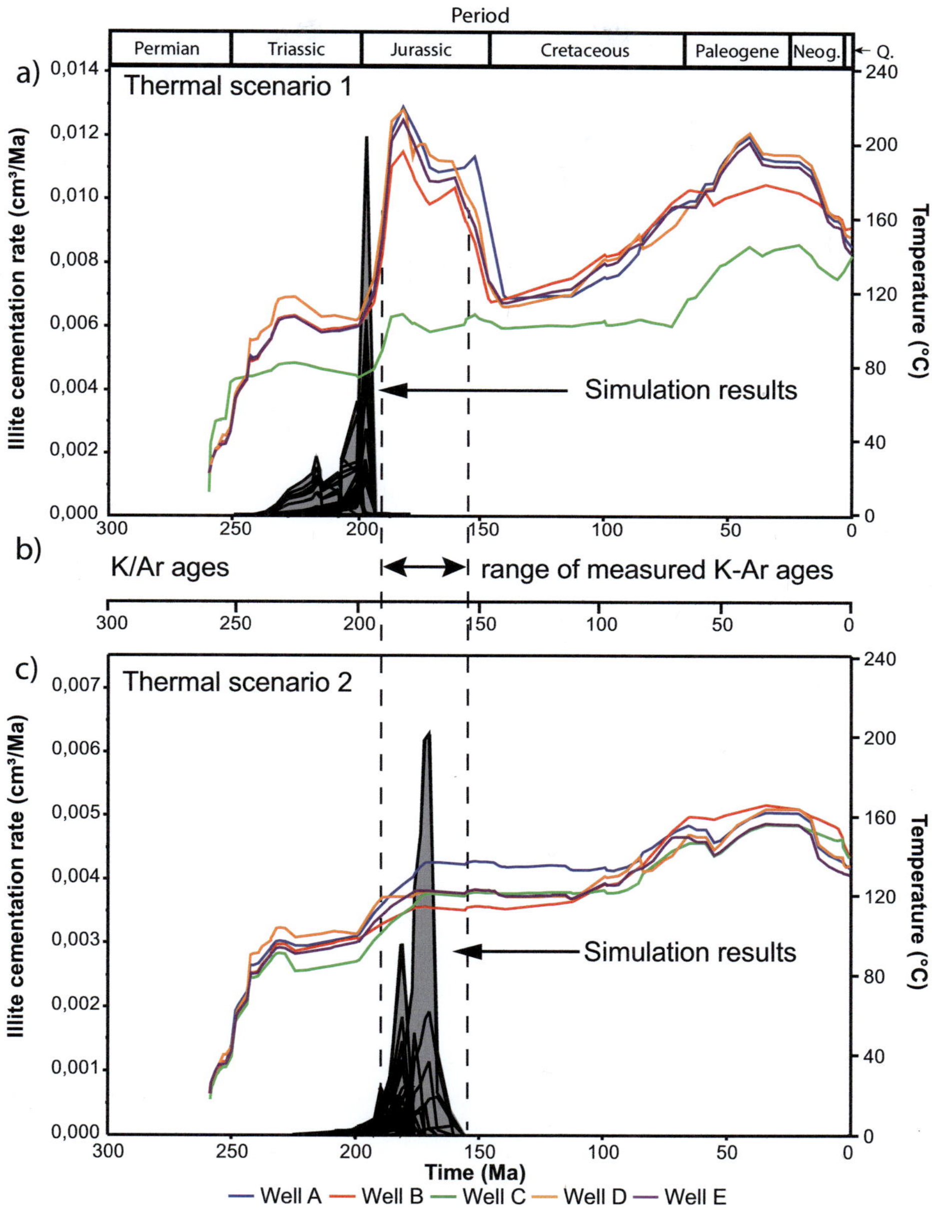

Figure 2.7: (a) Comparison of modeled rate of illite formation (black lines) for thermal scenario 1 of wells A-E (colored lines). The grey filled region marked as "simulation results" shows the modeled volumetric rate of illite formation through time. Note that the results for this scenario show illite formation beginning prior to the oldest K/Ar date. (b) Measured K/Ar ages of illites from the North German Basin from Zwingmann et al. (1998). (c) The illite simulation results for thermal scenario 2 predicts illite formation (black lines) within the range of measured illite ages for each well. The thermal reconstruction for wells A-E are given as the colored lines.

2.5.4 Porosity, Permeability, and Intergranular Volume

Given that thermal scenario 2 is far more consistent with observed quartz and illite cement volumes and characteristics, we used this scenario as a basis for developing a general reservoir quality prediction model. This model provides very good performance for total porosity, permeability, and intergranular volume (Fig. 2.8). The calculated porosities are within 4 vol% of the measured values that range from 1 to 16 vol% (Fig 2.8 a), except one porosity measurement from well A. The permeability values which range from 0.1 to 100 mD (Fig. 2.8 b), are similarly well matched with only two lower permeability predictions being in error by more than a factor of ten. The measured IGV values, which range from 16 to 37 %, are generally matched within 4 vol% by the simulations (Fig. 2.8 c).

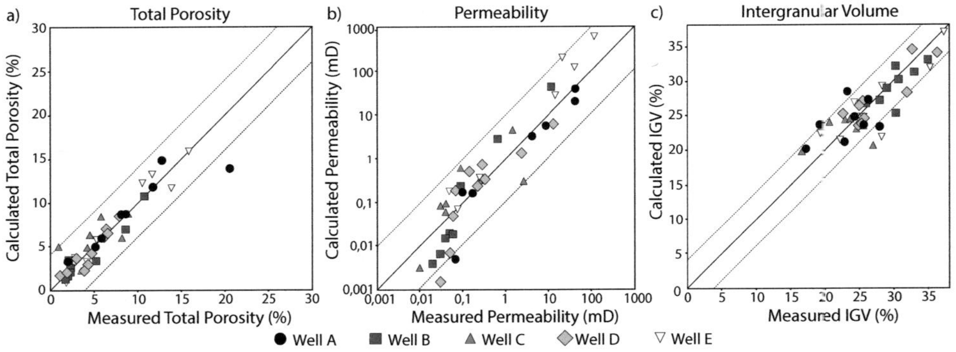

Figure 2.8: Calibration plots based on burial scenario 2 showing measured values on the x-axis and the model predicted values for the present-day on the y-axis. (a) Total porosities. (b) Permeabilities. (c) Intergranular volumes.

2.6 Discussion and conclusions

Although thermal scenario 1 provides an acceptable match to measured temperature and vitrinite reflectance data, it leads to overprediction of quartz cement abundance for wells B, D and E and underprediction of quartz cement from well C when a single set of kinetic parameters are used for samples from all wells. These results suggest that the scenario 1 thermal exposure is greater than indicated by quartz cement abundances for scenario 1 for wells A, B, D and E and lower for well C. Additionally, this thermal scenario leads to predicted timing of illite formation that pre-dates K/Ar dates of illite from the region.

In light of the poor match with quartz cement and illite data we adapted heat flow scenario 2 from heat flow reconstructions from Schwarzer and Littke (2007) and Uffmann and Littke (2011), which contrasts with scenario 1 in that the heat flow increase during the

Jurassic is omitted. Typical continental heat flow values around 55-60 $\frac{mW}{m^2}$ since the Jurassic sufficiently model the maturation of organic matter in this part of the North German Basin. The additional rifting phases in the North German Basin do not lead to an increase in heat flow in the basin models from which the heat flow was derived. Thermal histories from this scenario have an equivalent quality of fit to present-day temperature and vitrinite reflectance data as heat flow scenario 1 (Fig. 2.3) while also providing far superior performance for reconstructing quartz cement abundances and the timing of illite formation (Fig. 2.7).

Our results illustrate the utility of using diagenetic models for constraining thermal histories. Although quartz cementation rates are strongly affected by temperature, the overall abundance of quartz is far more sensitive to elapsed time compared to organic paleothermal indicators such as vitrinite reflectance (Awwiller and Summa, 1997; Lander et al., 1997a; Lander et al., 1997b; Perez et al., 1999). Recent publications highlight the accuracy of both quartz cementation kinetic models (Harwood et al., 2013) and vitrinite reflectance kinetic models (Le Bayon et al., 2012).

The kinetic model for fibrous illite formation is comparable to that of quartz with respect to the impact of both time and temperature. However, while quartz overgrowth cementation is a process that appears to continue as long as nucleation surface area is available, the illitization reaction stops once kaolinite and potassium (derived in this study from detrital K-feldspar alteration) reactants are consumed. Consequently, this reaction has the potential to provide useful information about absolute time that may be tested by comparison with radiometric dates (Lander and Bonnell, 2010). In this study the final peak illite cementation rates matched illite K/Ar dates that correspond to 190 – 155 Ma when temperatures were significantly below maximum values that were reached later in the thermal history.

The implications of improving constraints on thermal histories are of enormous importance not only in the study area but in many, if not most, hydrocarbon provinces. Thermal history is a primary control on the timing of hydrocarbon generation and oil-to-gas cracking, which in turn impacts hydrocarbon migration and the likely fluid types in undrilled reservoirs. Thermal history also clearly is a direct control on diagenetic processes of fundamental importance for reservoir quality. Quartz cementation and fibrous illite crystal growth kinetics provide uniquely powerful means for constraining thermal reconstructions when used in concert with well-established organic paleothermometers and petroleum systems modeling methods.

3 Cementation and structural diagenesis of fluvio-eolian Permian Rotliegend sandstones, North England

3.1 Abstract

Cementation and reservoir quality in sandstones are closely linked to grain coating phases and diagenetic alteration. It is widely known that grain rims consisting of illite/smectite stained with iron oxides and hydroxides are able to preserve large amounts of porosity by preventing the formation of syntaxial quartz overgrowth cement. The Penrith Sandstone Formation was chosen as an analog for other Rotliegend reservoirs. The half-Graben basin exposes the arid eolian depositional environments from alluvial conglomerates, via mixed fluvial-eolian deposits, to pure dune deposits. Differences of grain coating characteristics can be linked to grain size. In well cemented sandstone samples, the grain coatings are better developed in finer grained laminae. Coarser grained laminae expose large amounts of syntaxial quartz overgrowth. Structural features include normal faults and strike-slip faults of different ages. Diagenetic alterations around faults allow the delineation of relative temporal relations, revealing at least two generations of deformation band formation associated with normal faulting. In the Vale of Eden succession one normal faulting event postdates burial diagenetic quartz cementation as evident by fault focused fluid flow and associated leaching of iron and absence of quartz overgrowth.

3.2 Introduction

Cementation and chemical compaction are the two main factors controlling reservoir quality in deeply buried sandstone reservoirs (e.g. Paxton et al., 2002; Taylor et al., 2010; Taylor et al., 2015). Additionally, reservoir quality can be affected by structural compartmentalization by faulting and fault-associated processes such as deformation bands (Antonellini and Aydin, 1994; Fisher and Knipe, 2001; Fossen and Bale, 2007; Tueckmantel et al., 2012). In sandstones, reservoir quality can be affected by depositional and early diagenetic processes such as grain coatings inhibiting quartz overgrowth cementation (Ajdukiewicz et al., 2010; Taylor et al., 2010; Ajdukiewicz and Larese, 2012). One of the depositional environments affected by grain coatings are eolian dune deposits which can preserve considerable amounts of porosity during diagenesis and deep burial (Ajdukiewicz et al., 2010).

Clay coating phases are thought to form from precursor clay mineral phases deposited on grain surfaces during deposition and early diagenesis (Ajdukiewicz and Larese, 2012). During burial diagenesis, these precursor minerals, mostly smectite, can transform to other clay minerals such as tangential or radial illite or chlorite (Aagaard et al., 2000; Worden and Morad, 2002). An important factor during the emplacement of clay minerals into the sediment body seems to be a periodical influence of meteoric waters as well as more arid phases (Salem et al., 2000; Esch et al., 2008; Ajdukiewicz et al., 2010). The climatic conditions and the supply of meteoric waters are therefore discussed to have a major impact on early diagenetic clay coat formation (Esch et al., 2008; Ajdukiewicz et al., 2010). To predict reservoir quality variations on a reservoir scale, reservoir analogs are studied to provide the areal extent covering a reservoir scale.

Deformation bands are zones of localized deformation found in granular media such as siliciclastic rocks and limestones (e.g. Antonellini and Aydin, 1994; Antonellini and Aydin, 1995; Fossen et al., 2007; Wennberg et al., 2013). The formation of deformation bands is linked to faulting processes, and they can form from consolidated and unconsolidated sandstones as precursors or concurrent to fault localization (Fowles and Burley, 1994; Antonellini and Aydin, 1995; Rotevatn et al., 2007; Johansen and Fossen, 2008). Deformation bands generally reduce permeability (Fossen and Bale, 2007; Tueckmantel et al., 2012; Busch et al., 2015), but permeability may locally be enhanced parallel to the strike of the deformation bands in the studied area (Fowles and Burley, 1994). During fault focused fluid migration the fault's damage zone and the surrounding lithology can be altered, as evident from color alterations of faulted red-bed sediments (Eichhubl et al., 2004; Parry et al., 2004; Eichhubl and Flodin, 2005). This alteration can be used to infer the diagenetic history and to delineate fault-related alteration of coating mineral phases.

In this study, we present petrographic data from a reservoir analog of eolian Rotliegend rocks, where zones of preferential quartz cementation were previously reported (Arthurton and Wadge, 1981; Turner et al., 1995). This lithology can be seen analogously to some of the Northern European onshore as well as offshore plays. Notable areas are the Permian Central European Basin System and the East Irish Sea Basin. Additionally, we compare QEMSCAN® analyses to standard petrographic analyses and establish a paragenetic sequence and derive systematic trends within the lithology and in dependence of faults. The correlation to the timing of faulting is used to confine the phases of differing diagenetic alterations. Especially the occurrence and possible alteration of clay coats is in the focus of this work.

3.3 Geological setting

The outcrops of Permian strata in the Vale of Eden half-graben, Cumbria, UK, cover an area of approximately 48 x 6 km (Macchi, 1981; Turner et al., 1995) (Fig. 3.1 a). Permian strata in the region are combined in the Appleby Group, consisting of the eolian Penrith Sandstone Formation and basal alluvial deposits, including fluvial deposits locally referred to as the Brockram facies (Arthurton and Wadge, 1981; Macchi, 1981; Hughes, 2003). Stratigraphic variations can be studied due to the regional gentle dip of 5° towards the north-east, lateral variations can be studied along the strike of the formation in several quarries and exposures (Fig. 3.1 b). The dip of the strata changes towards the Pennine Fault System to a dip of approximately 20° towards the south-west (Fig. 3.1 b) (Arthurton and Wadge, 1981). They unconformably overlie Upper Carboniferous siliciclastics in the northern part of the studied area and Lower Carboniferous limestones in the south (Arthurton and Wadge, 1981; Macchi, 1981; Turner et al., 1995) (Fig. 3.1 b, c). The formation is conformably overlain by the Zechstein deposits consisting of shales, evaporites, and dolomite, which are combined in the Eden Shales Formation (Arthurton and Wadge, 1981). Triassic deposits in the region are present as the St. Bees Sandstone of the Sherwood Sandstone Group (Hughes, 2003). The half-graben is bound by the Dent Fault Line to the South and the Pennine Fault System to the north-east (Underhill et al., 1988; Fowles and Burley, 1994) (Fig. 3.1 a). The half-graben formed during early Permian rifting and subsidence was actively ongoing through to the Cretaceous to a maximum depth of approximately 3500 m (Turner et al., 1995). Reactivation and basin inversion from the late Cretaceous through to the present uplifted the strata to its present position (Turner et al., 1995). The structural inventory also includes deformation bands associated to normal faulting (Fowles and Burley, 1994; Guo et al., 2009)

Turner et al. (1995) proposed a model for cementation of the formation. The localized cementation is linked to the dip of the formation during burial and the leveling of the paleowater table with the saline, more quartz-rich fluid phase (Turner et al., 1995) collecting below an unconformity, resulting in localized abundance of quartz cement. The most likely source of silica was discussed to be alkaline, saline fluids from the Upper Permian (Zechstein) Eden Shales Formation (Turner et al., 1995). Such formation waters are able to support a higher saturation of silica in solution than the more acidic formation waters of Upper Carboniferous strata, containing sandstones, shales, and coal seams (Pennine Coal Measures Group and Millstone Grit) (Turner et al., 1995). The source of the silica-bearing solutions is additionally supported by saline fluid inclusions in quartz overgrowth cements measured in samples from the Penrith Sandstone Formation (Turner et al., 1995).

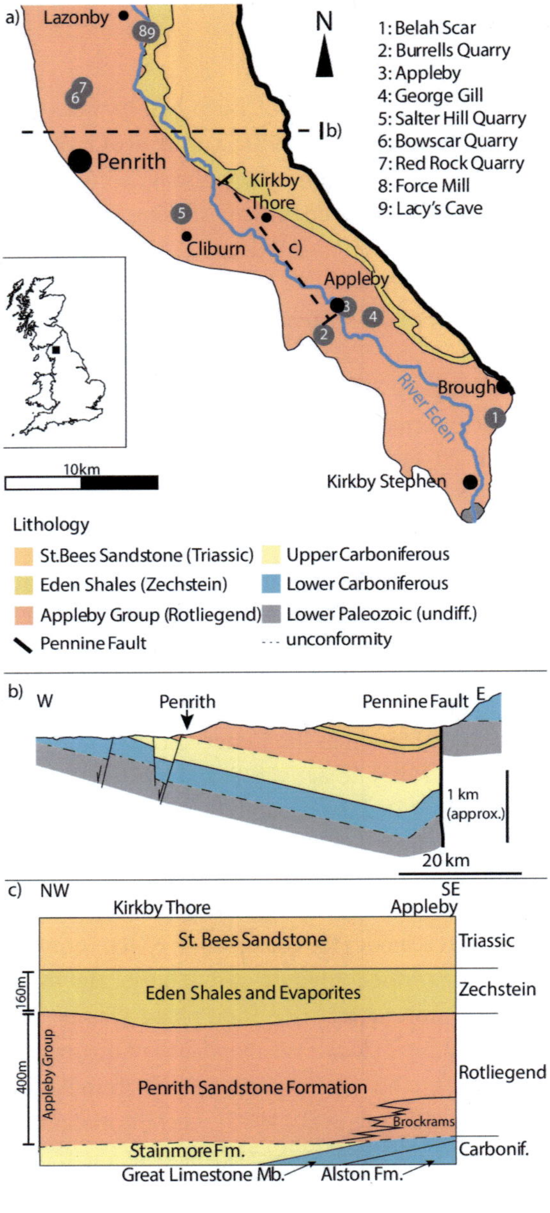

Figure 3.1: a) Overview of the studied area in the Vale of Eden half-graben, Cumbria, UK. The locations of the nine studied exposures are highlighted (1-9). The lines of profile have been indicated using dashed lines. b) Schematic profile of the Vale of Eden half-graben north of Penrith. The Permo-Triassic units of the Vale of Eden gently dip towards the north-east (redrawn and modified from Turner et al. (1995), vertical scale is given separately). c) The NW-SE profile reveals the unconformity between Carboniferous and Permian strata and regional extent of the basal alluvial fan Brockram facies (modified from Hughes (2003)). The Eden Shales and St. Bees Sandstone have been projected onto the cross section. The color for the lithological units are based on Hughes (2003).

3.4 Materials and Methods

Oriented samples were collected from disused and active quarries (Red Rock, Bowscar, Salter Hill). At sites of special scientific interest (SSSI), only rocks fallen from the outcrop were analyzed. No samples were collected inside Lacy's Cave (Fig. 3.1a). Surrounding the caves and on the opposite bank of River Eden, a few samples were collected. Cementation of the samples has mainly been studied on thin sections from multiple locations in the Vale of Eden half-graben. Thin sections were stained with a blue-dyed epoxy-resin to visualize porosity and prepared to be 30 µm thick. Point counting was performed with 300 counts in total with a Pelcon Point Counter mounted on a Zeiss Axio Lab microscope. General petrographic analysis was performed with a Leica DMLP microscope fitted with an Invenio 5DII camera controlled by DeltaPix Insight software. For each sample the coat coverage was evaluated based on 50 grains per sample. The coat coverage is the relative amount of grain surface covered by a grain coating phase which is in contact with the intergranular volume (IGV). Grain contacts are omitted in this evaluation but yield information about a possible detrital origin of the grain coating phase. Grain sizes and sorting characteristics were derived from measurements of the long axis on at least 100 grains per sample on a grid adjusted to the maximum grain size to retrieve area weighted results.

Two polished thick sections (100 µm) were prepared and coated with carbon for QEMSCAN® analyses. QEMSCAN® analyses were applied to visualize the distribution of minerals within selected samples. QEMSCAN® is a semi-automated mineralogical tool based on an SEM platform. Supported by fast acquisition energy dispersive X-ray spectroscopy (EDX; DualXFlash SDD Brucker) and back-scattered electron/secondary electron (BSE/SE) imaging detectors, QEMSCAN® has the advantage to scan large sample surfaces at high resolutions. EDX spectra and BSE grey-level intensities were recorded for each sampling point and converted to a pixel information. This combination to assessable pixel properties is used for phase/mineral interpretation by adjustable purpose-built species interpretation profiles (SIP). Pixels of the same phase sharing edges are handled as particles. By image analysis, these particles are surveyed for geometrical (e.g. size, shape and associations) and chemical (elemental distribution and mapping) properties. Volumetric distributions of phases are gathered from the surface coverage of the specific phases to generate modal compositions.

Analyses of the carbon coated samples were performed with a Quanta 650-F QEMSCAN® (FEI) platform with a point-spacing of 5 µm. The electron beam was accelerated by 15 kV and the sample current was held at 10 nA and both automatically checked and adjusted by a two-hour cycle during the whole measurement. For each pixel EDX spectra was recorded until 2000 counts were gathered. The working distance (sample surface to pole piece) was set to 13 mm.

3.5 Results

3.5.1 Exposure Characteristics

The outcrops comprise SSSI at Belah Scar (BS, Fig. 3.2 b, #1 in Fig. 3.1 a) and Burrells Quarry (#2 in Fig. 3.1 a), exposing alluvial fan and fluvial deposits. Burrells Quarry represents the lowermost part of the Appleby Group, covering alluvial fan deposits unconformably overlying the Carboniferous limestones. The outcrop at George Gill (Fig. 3.2 e, #4 in Fig. 3.1 a) is also a SSSI, comprising the top of the Penrith Sandstone Formation with intensive faulting and fault associated deformation band formation. For the outcrops at Burrells, a roadcut in Appleby-in-Westmorland (Fig. 3.2 c, #3 in Fig. 3.1 a), and at George Gill, no petrographic samples were taken. The outcrops at George Gill also exhibit small channelized deposits including coarser grained (coarse sand) lag deposits. The active quarries at Bowscar Road (Fig. 3.2 a, #6 + 7 in Fig. 3.1 a), north of Penrith, expose dune bodies from a central part of the Penrith Sandstone Formation. A disused quarry at Salter Hill (SH, #5 in Fig. 3.1 a), north of Cliburn, features laterally continuous and unfaulted dune cross-beds (Fig. 3.2 d). Faulting and deformation bands are well exposed in and around Lacy's Cave (LC, #9 in Fig. 3.1 a) and Force Mill (FM, #8 in Fig. 3.1 a) (Fig. 3.2 f). Leaching of originally red sandstones occurs around some faults in the lithology, including small strike slip faults in Red Rock Quarry (RR, #7 in Fig. 3.1 a) (Fig. 3.2 g) and Bowscar Quarry (BQ, #6 in Fig. 3.1 a). The normal faults and deformation bands locally obstruct the identification of sedimentary texture. Away from deformed parts, the sedimentary texture can be identified. Cross-bedding in eolian sandstones cannot be reconstructed within Lacy's Cave. However, underneath and to the north the cross bedding of dunes is well exposed. The sedimentary textures in Force Mill include large-scale dune cross bedding.

Faulting is observed in most exposures and faults are occasionally associated with color alterations of the surrounding host rock (Fig. 3.2 f, g).

Figure 3.2: Images of fluvio-eolian Rotliegend outcrops in the Vale of Eden. a) Overview of Bowscar Quarry featuring extensive large scale dune deposits. b) Outcrop at Belah Scar featuring channelized conglomerate layers (blue dashed line) interbedded with sheet flood deposits and fluvial channels. c) Roadcut in Appleby-in-Westmorland exposing interbedded eolian dune deposits, including normal faulting (black dashed line) and deformation band formation (blue dashed line). d) Disused quarry at Salter Hill north of Cliburn, featuring interbedded large dune deposits. e) Intensely faulted Penrith Sandstone at George Gill, including closely spaced deformation band formation. f) Greyish-white zone associated to deformation band clusters outside of Lacy's Caves. The undisturbed red color of the host rock is maintained at some distance to the cluster or single bands. g) Bleaching is also prominently observed around small strike slip faults with an undeterminable offset. Shear indicators like slickensides point towards both a sinistral and dextral movement.

3.5.2 Petrography

The fluvial samples from Belah Scar (Brockrams and Penrith Sandstone Formation) have an average grain size ranging from 0.1 and 0.2 mm and show a weak lamination and moderate sorting. The conglomeratic alluvial sample features clasts above 2 cm in diameter in a fine to medium sand matrix and an overall moderately poor sorting. The eolian samples from Bowscar, Red Rock, Salter Hill, Force Mill, and Lacy's Cave (all Penrith Sandstone Formation) feature average grain sizes from 0.15 to 0.47 mm. The sorting is moderate to moderately well, occasionally showing bimodal laminations.

Detrital components of the fluvio-eolian Rotliegend Penrith Sandstone and the alluvial Rotliegend Brockrams were determined by point counting 18 thin sections, giving quartz grains 88 – 96 %, feldspar grains including plagioclase 4 – 9 % and minor amounts of rock fragments of 0.3 – 6 % (Tab. 3.1). The only exception is the conglomerate sample from Belah Scar (BS_congl) with only 83 % quartz, 3 % feldspars, and 14 % rock fragments the latter being mainly limestone clasts. The sandstones can be classified according to Folk (1980) as arenites to subarkoses for the fluvial and eolian sandstones, and lithic subarkose for the alluvial clastic sample from Belah Scar (Fig. 3.3).

Quartz grains generally exhibit an iron-rich coating on the surface which can be divided into two groups of pigmented hematite dust rims and red-stained illite coatings (Fig. 3.4). The amount of syntaxial quartz cement varies from 0.3 - 20.7 % by volume in undeformed rock samples (Tab. 3.1).

The channel deposits (BS_1 and 2) exhibit illite coatings which are stained with iron oxides/-hydroxides (Fig. 3.4 a, b). These samples do only exhibit small amounts of quartz cements of 0.3 % by volume.

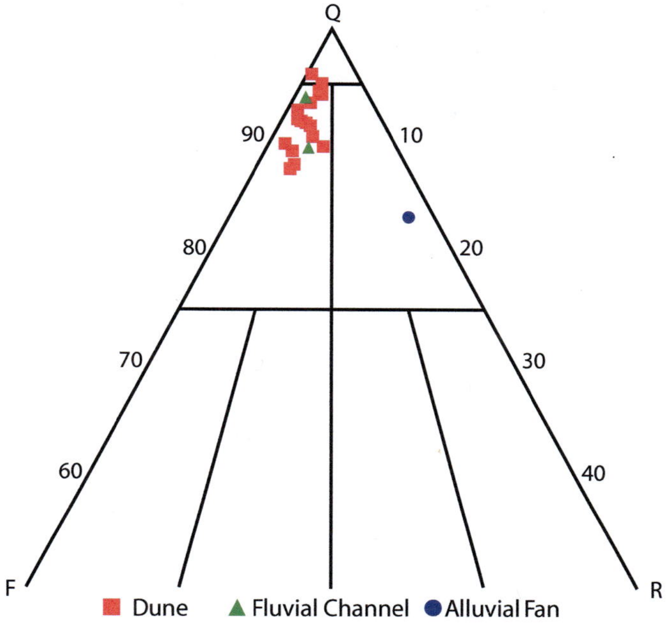

Figure 3.3: Ternary QFR diagram showing the composition of the studied samples from the Vale of Eden.

In eolian samples, well developed illite coatings were observed in samples from Force Mill (FM) in the northern part of the study area (Fig. 3.4 e, f). These samples also show low amounts of quartz cement with 1 – 7.3 % by volume. In areas where a beige color alteration is observed, the illite is locally unstained (see chapter 3.6.3). Samples from the three quarries north and south of Penrith (BQ, RR, SH) in the northern part of the study area are characterized by the largest amount of quartz cement of all samples with 11 to 20.7 % by volume (Fig. 3.5 a-d). Where a locally continuous layer of hematite or illite is present on quartz grains, syntaxial overgrowths are not present (arrows in Fig 3.5 c, d).

Prominent rock fragments include silicic and feldspar rich volcanic fragments, shale rock fragments, sandstone and siltstone rock fragments, and metamorphic rock fragments in the other samples. The conglomerate (BS_congl, Fig 3.4 c, d) is characterized by large limestone clasts and coarse sand-sized to silt-sized grains cemented by calcite.

Table 3.1: Compilation of point count results for the samples from the studied area including the evaluation of clay coat coverage and additional sample information, i.e. the depositional environment, mean grain size, and sorting.

Detritus	BS_1	BS_2	BS_congl	BQ_1	BQ_2	BQ_3	BQ_4	FM_1	FM_2	FM_3	FM_Alt	LC_1	LC_2	SH_1	RR_1	RR_2	RR_3	RR_4	RR_5
Quartz (incl. undulose and polycrystalline)	70.7	69.3	64.0	68.3	64.7	71.3	60.0	78.7	67.0	65.7	72.1	71.3	76.0	63.7	65.0	66.3	73.0	69.3	70.7
K-feldspar	4.0	5.3	1.3	2.7	6.3	5.0	6.0	4.0	6.0	4.3	3.0	4.3	2.7	4.0	4.7	6.0	2.7	4.3	4.7
Plagioclase	0.0	0.3	0.3	0.0	0.0	0.0	0.3	0.3	0.3	0.0	0.0	0.3	0.0	0.3	0.0	0.7	0.0	0.0	0.0
Clay matrix	4.0	2.0	1.3	0.0	0.0	0.3	0.0	0.3	0.0	0.0	0.0	0.0	0.0	0.3	0.0	0.0	0.0	0.0	0.3
Mica	0.0	0.0	0.3	0.0	0.0	0.0	0.0	0.0	0.0	0.0	0.0	0.3	0.0	0.0	0.0	0.3	0.0	0.7	0.3
Hematite	0.0	2.0	0.3	0.0	0.0	0.0	0.0	0.0	0.7	0.7	0.0	0.3	0.0	0.0	0.0	0.0	0.0	0.0	0.3
Accessory minerals (titanite, tourmaline, zircon)	0.0	0.7	0.0	0.0	0.0	0.0	0.0	0.0	0.0	0.0	1.0	0.0	0.0	0.0	0.0	0.0	0.0	0.0	0.0
Shale rock fragment	0.0	0.3	0.0	0.0	0.3	0.0	0.0	0.3	0.0	0.0	0.0	0.0	0.0	0.7	0.0	0.0	0.0	0.0	0.0
Sandstone rock fragment	0.0	0.0	0.3	0.0	1.0	0.3	0.0	0.0	0.3	0.0	0.3	0.3	0.3	0.3	0.3	0.0	1.0	1.0	0.3
Volcanic rock fragment	0.7	2.0	1.0	0.3	0.3	0.3	2.0	1.0	1.3	1.0	0.7	1.3	1.0	2.0	1.7	1.0	0.7	1.0	0.7
Plutonic rock fragment	0.0	0.0	0.0	0.0	0.0	0.0	0.3	0.0	0.7	0.7	0.0	0.3	0.0	0.0	0.0	0.0	0.3	0.3	0.7
Metamorphic rock fragment	0.0	0.3	0.0	0.0	0.0	0.0	0.0	0.0	0.7	0.0	0.3	0.0	0.0	0.3	0.3	0.0	0.0	0.0	0.0
Authigenic phases																			
Quartz cement	0.3	0.3	0.0	20.7	11.0	11.3	18.0	1.0	11.7	7.0	2.3	5.0	7.3	16.5	13.3	14.7	13.7	12.3	5.7
K-feldspar cement	0.0	0.0	0.0	0.3	0.0	0.0	0.0	0.0	0.3	0.7	0.0	0.0	1.0	0.3	0.0	0.0	0.7	0.3	0.0
Illite pore lining	3.3	1.7	1.0	0.0	1.3	0.3	0.3	1.3	1.3	1.7	4.7	2.3	2.0	1.7	2.0	0.0	0.3	0.0	0.7
Illite pore filling	1.7	1.0	0.0	0.0	0.3	0.0	0.0	1.0	0.0	0.0	3.3	0.3	0.7	0.7	0.0	2.0	0.0	0.7	0.0
illite replaces K-feldspar	3.7	3.0	0.7	1.0	1.7	0.0	1.7	0.7	2.7	2.0	1.0	2.3	2.7	2.3	4.3	4.3	1.7	2.3	3.3
Kaolinite	0.0	0.0	0.0	1.3	1.3	1.3	0.7	0.3	0.0	0.3	0.7	0.3	0.0	1.3	1.3	1.0	1.0	0.7	0.7
Calcite	0.3	2.7	14.3	0.0	0.0	0.0	0.0	0.0	0.0	0.0	0.3	0.0	0.0	0.0	0.0	0.0	0.0	0.0	0.0
Siderite	0.0	0.0	0.0	0.0	0.0	0.0	0.0	0.0	0.0	0.0	0.3	0.0	0.0	0.0	0.0	0.0	0.0	0.0	0.0
Anhydrite	0.0	0.0	0.0	0.0	0.0	0.0	0.0	0.0	0.0	0.0	0.3	0.0	0.0	0.0	0.0	0.0	0.0	0.0	0.3
Iron oxide pore lining	0.3	0.7	0.0	0.0	0.7	0.3	0.3	0.3	2.0	1.7	0.0	0.7	0.3	0.0	0.7	0.0	0.7	0.0	1.0
Iron oxide pore filling	0.0	0.7	0.0	0.0	1.0	0.0	1.0	0.3	0.0	0.0	0.3	0.0	0.0	0.3	1.0	0.3	0.0	0.0	1.0
Calcite replaces K-feldspar	0.0	0.0	3.0	0.0	0.0	0.0	0.0	0.0	0.0	0.0	0.3	0.0	0.0	0.0	0.0	0.0	0.0	0.0	0.0
Porosity																			
Porosity	11.0	7.3	1.7	5.0	8.0	8.7	9.3	9.3	4.7	13.7	8.6	9.7	5.0	4.6	3.7	1.7	4.0	7.0	8.0
Secondary porosity	0.0	0.3	0.3	0.3	2.0	0.7	0.0	1.0	0.0	0.7	0.3	0.7	0.7	0.7	1.7	1.7	0.3	0.0	1.3
Grain coat coverage																			
Grain coat coverage (%)	70.06	79.9	n.a.	13.6	25.52	13.9	21.16	95.32	37.12	61.74	83.88	70.06	74.48	19.1	31.18	20.3	15.18	19.8	38.18
Additional sample information																			
Depositional environment	fluv. ch.	fluv. ch.	all. fan	dune	dune	dune	dune	dune	dune	dune	dune	dune	dune	dune	dune	dune	dune	dune	dune
Mean grain size (mm)	0.23	0.10	n.a.	0.47	0.36	0.41	0.15	0.34	0.18	0.19	0.17	0.35	0.38	0.45	0.29	0.32	0.19	0.20	0.18
Sorting	ms	ms		ms	mws	ms	ms	mws	ms	ms	ms	mws	ms	mws	mws	mws	ms	ms	ms

mws: moderately well sorted ms: moderately sorted

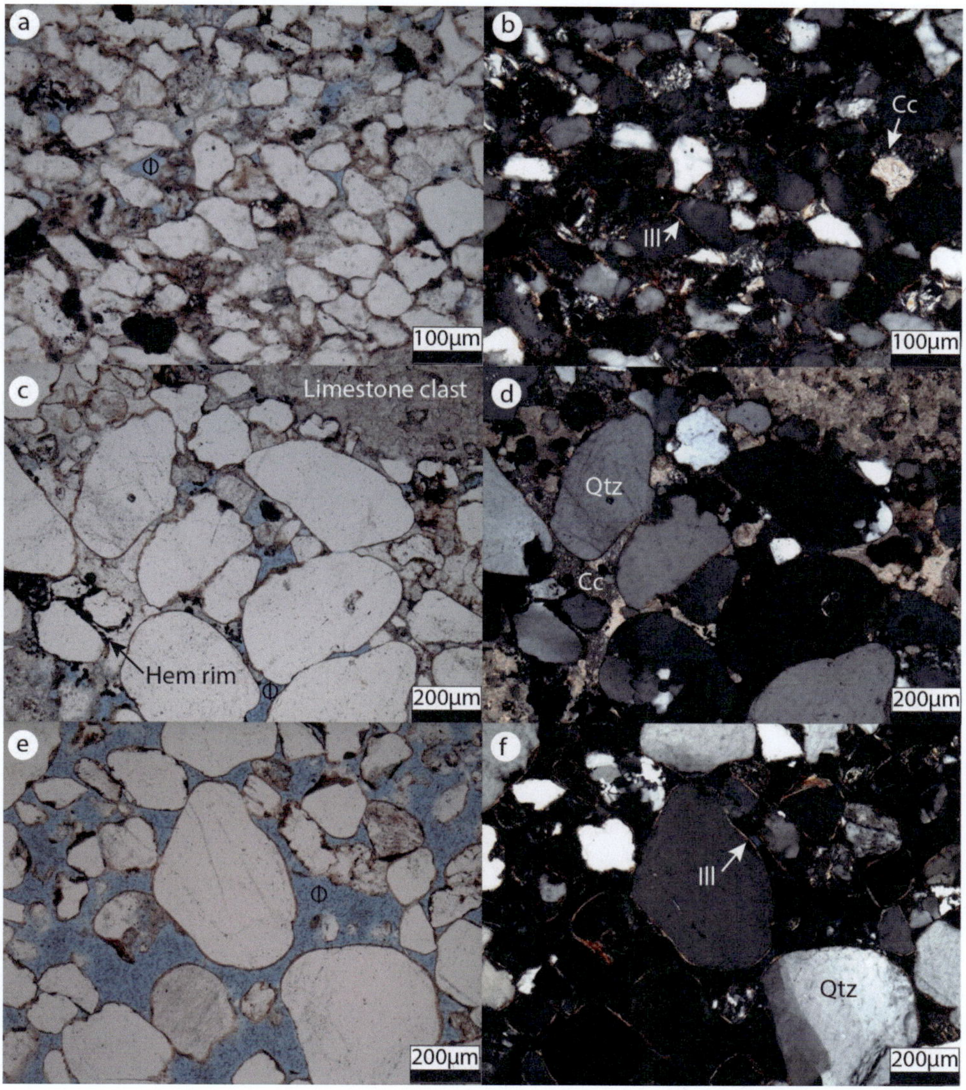

Figure 3.4: Thin section images from Belah Scar and Force Mill. a, b) Thick brown rims consisting mainly of iron oxides/-hydroxides and illite as visible in ppl and xpl from a channel deposit (BS_1) from Belah Scar. Some of the pore space is cemented by calcite, and the illitization of feldspars favors the development of a pseudomatrix texture. c, d) Well developed brown rims surrounding the sand-sized grains in this sample from one of the conglomeratic layers (BS_congl) from Belah Scar, featuring limestone clasts (top right and bottom left) as well as quartz grains in a carbonate cement matrix. The effect of coatings in this part of the outcrop is negligible due to the absence of quartz overgrowth cements. e, f) Intact stained tangential illite coatings are present and inhibit quartz overgrowth cementation in sample FM_1 from Force Mill. Legend: Φ – blue-dyed pore space, Cc – calcite cement, Ill – illite, Hem – hematite, Qtz – quartz.

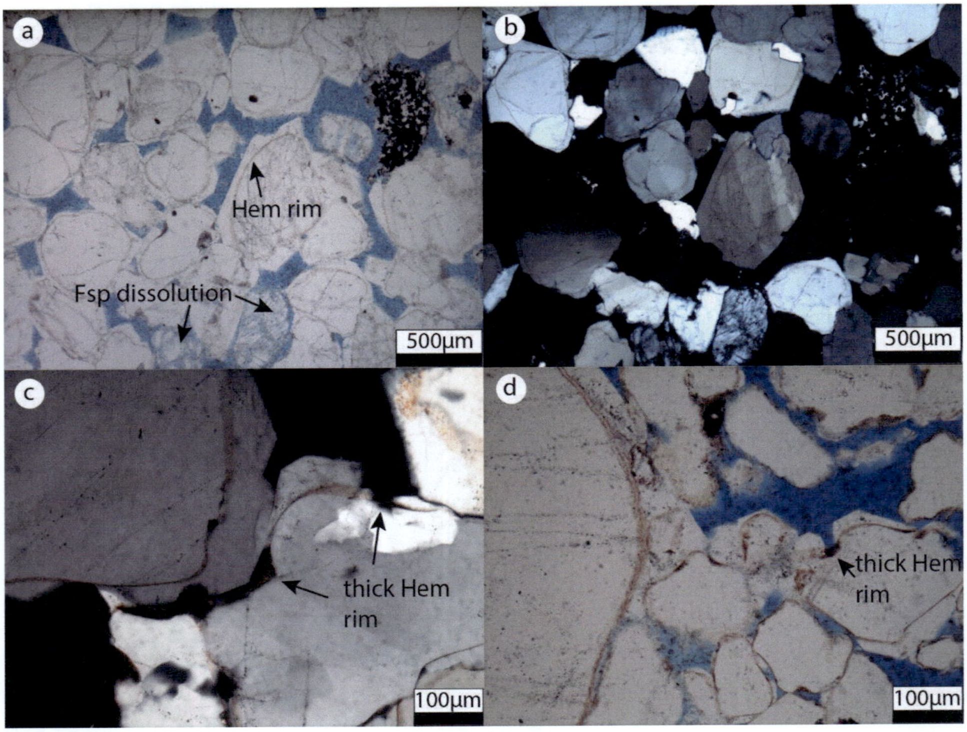

Figure 3.5: Eolian dune samples from Bowscar Quarry. a, b) Pigmented hematite dust rims outlining the original grain surface without illite coating the grain surface (BQ_1). c, d) Thicker iron oxides/-hydroxides linings and occasionally illite inhibit quartz cementation (BQ_1). The thicker continuous iron oxides/-hydroxides are accumulated in surface roughnesses and indentations. Hem: hematite, Fsp: K-feldspar.

Feldspars are often dissolved or altered and either replaced by kaolinite or illite (Fig. 3.6 b-d). Skeletal frameworks of dissolved feldspars are often surrounded by the replacive mineral. Occasionally the feldspars are dissolved without any replacive phases in the vicinity. The dust rim outlining the old grain boundary is often preserved during any of the alterations (Fig. 3.6 b). Upon reconstruction of the feldspars before alteration, the detrital composition changes towards a more feldspar rich composition, with all sandstone samples in the field of subarkoses with a maximum feldspar content of 23 % in a Quartz-Feldspar-Rock fragment diagram. Most of the present kaolinite is unaltered and illitization of kaolinite in the studied samples could be well observed in QEMSCAN® analyses and transmitted light microscopy (Fig. 3.6 e, f). Most feldspars are at least slightly illitized. Detrital mineral grains include in descending order iron oxide grains, titanite, muscovite, tourmaline, and zircon.

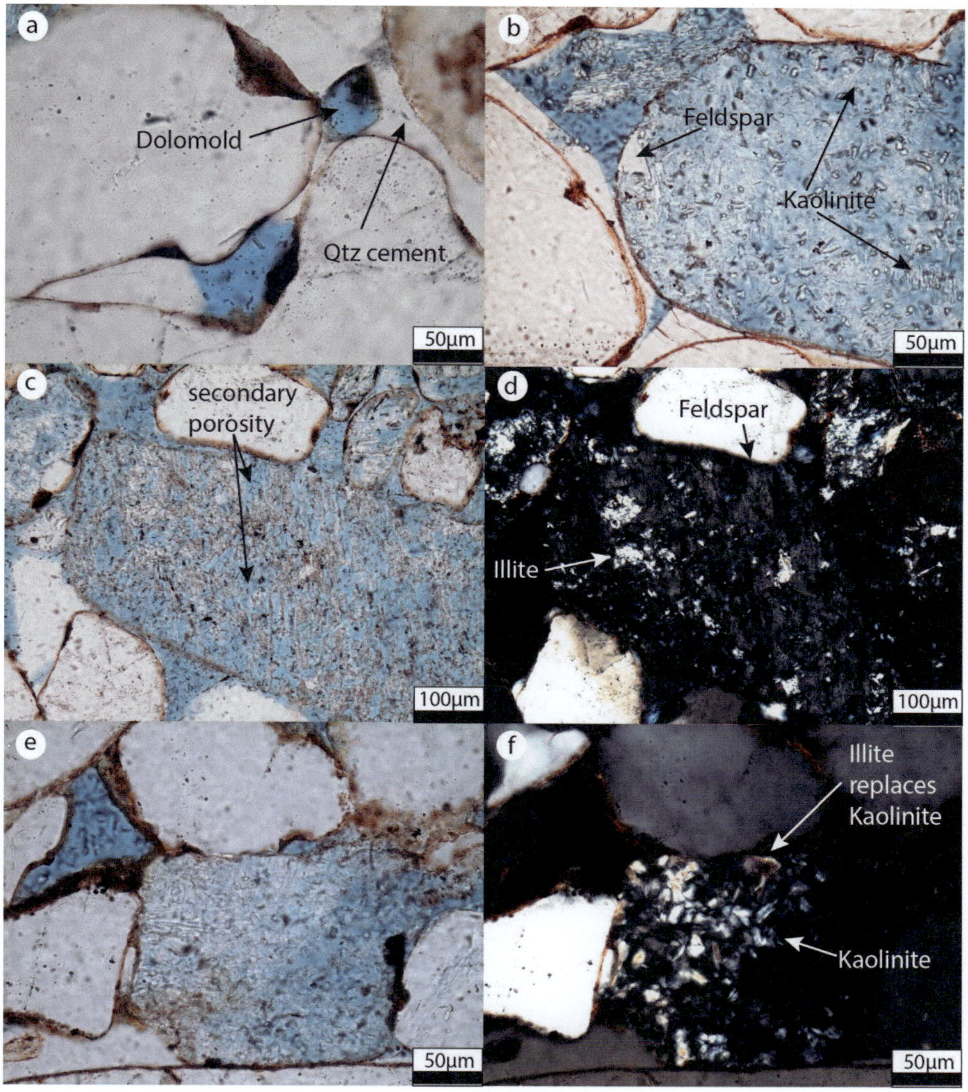

Figure 3.6: Prominent authigenic alterations in eolian samples. a) Dolomold enclosed within syntaxial quartz cements in sample BQ_2. b) Kaolinite replaces feldspar grain while the original grain boundary, marked by pigmented hematite rim, is still preserved is sample BQ_2. c, d) Illite replaces feldspar grain, additionally secondary porosity is generated during the dissolution in sample RR_1. e, f) Kaolinite replaced a detrital grain and is subsequently replaced by illite, still preserving the booklet texture in sample BQ_2. The original grain outline is still preserved, indicating that pervasive grain dissolution and replacive kaolinite formation occurred after significant compaction.

Carbonate and anhydrite cements in eolian samples (BQ, FM, LC, RR, SH) are only present in small quantities around 0.3 %. Carbonate cements include calcite, siderite, and dolomite and appear partly dissolved. Dissolved carbonate cements are evident from rhombic pores (dolomolds) enclosed within quartz overgrowth cements in the eolian dune samples (Fig. 3.6 a). In the fluvial and alluvial samples from Belah Scar, calcite represents the main cement phase which locally varies in abundance (Fig 3.4 a-d). Rarely the outcrop samples feature small amounts (0.3 %) of anhydrite, present as a grain framework stabilizing, pore-filling cement.

3.5.3 Grain coatings

The correlation of grain coat coverage to the amount of quartz cement for all studied samples emphasizes the strong impact of clay coats in the studied area (Fig. 3.7). The exception being the intensely calcite cemented fluvial channel samples featuring very low quartz cement volumes although the grain coating coverage is at 70 - 80 % (Fig. 3.7). Within dune sandstones, the grain coat coverage does not vary systematically (samples BQ_1 to 4). However, an increase of grain coat coverage from samples in quarries north (BQ, RR) and south of Penrith (SH) towards samples collected near the Zechstein contact at Force Mill and Lacy's Cave can be observed (see table 3.1). This effect is additionally documented by Waugh and Macchi (1988, their Fig. 17+18) where an unfaulted sample from George Gill, near the Zechstein contact, shows hematite and clay rims, while quartz cement is generally absent. The fluvial channel samples from Belah Scar (BS_1 and 2) exhibit relatively large grain coat coverages compared to the dune samples from the quarries in the northern part of the studied area. The pervasive carbonate cement from the alluvial sample from Belah Scar hindered the correct evaluation of grain coat coverage.

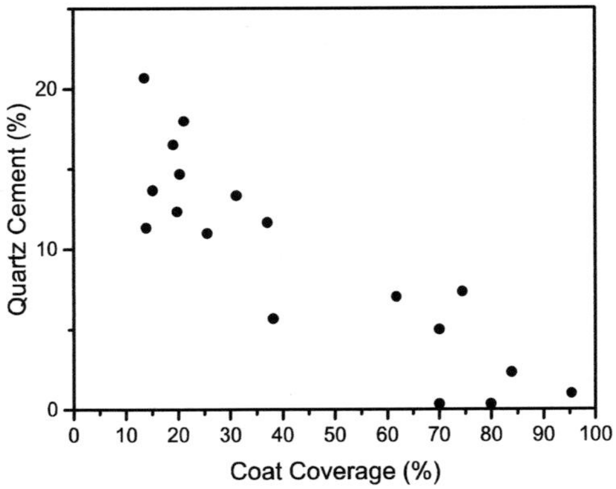

Figure 3.7: Correlation between point-counted quartz cement values and observed average grain coat coverage. The decrease in quartz cement correlates well with the grain coat coverage. It has to be noted that samples affected by pervasive carbonate cementation show the lowest quartz cement abundance.

3.5.4 QEMSCAN® results

QEMSCAN® analyses can highlight the distribution of mineral phases within the samples. Additionally, the grain coating characteristics were evaluated using QEMSCAN®.

The main part of detrital components of sample BQ_2 consists of quartz (pink, Fig. 3.8 b) with 87.1 %. The content of K-feldspar (blue-green) in the studied sample from Bowscar Quarry is at 6.5 %. The K-feldspars appear partly dissolved (Fig. 3.5 a, b and bottom in Fig. 3.8 b). Some feldspar grains are kaolinitized or illitized whereas others are partly dissolved. In the presented sample, the bulk illite content is 3.5 %. Kaolinite is sparse with 1.3 % while chlorite, often associated with illite and rock fragments, is rarely present with 0.7 %. This abundance at a mean particle size between 6.3 and 20 μm was not detected during point counting. Considering grain coatings, the fine sand-sized grains of the samples from Bowscar and Red Rock Quarry are characterized by a relative enrichment in illite as compared to the coarser grained sand fraction. The fine grained sand fraction (lines 3 and 5 in Fig. 3.8 a, b) contain 6.1±0.1 % illite (green, Fig. 3.8 b) whereas the coarser intervals (lines 1, 2, 4 and 6, Fig. 3.8 a, b), contain 2.2±0.7 %. Additional transmitted light petrographic analysis highlights the location of illite in the fine sand fraction as a grain rimming phase and does not occlude the pore space (Fig. 3.8 c-f).

Oxide phases detected during QEMSCAN® analyses are mainly hematite and rutile, which are part of the detrital inventory of the sediment. However, this method is not able to detect all pigmented hematite rims enclosed by later syntaxial quartz cement.

The mineralogical inventory of the QEMSCAN® analyses generally matches well with point count analyses (Tab. 3.2). Differences can only be seen for mineral phases with a low concentration within the sample. The threshold in this sample appears to be at 0.65 %. However, point counting was also able to identify hematite at 1.85 %, where QEMSCAN® analyses resulted in smaller abundances at 0.4 %. Rock fragments however, cannot be simply detected using QEMSCAN® analyses, whereas a petrographer can interpret the rock type based on mineralogical and textural parameters.

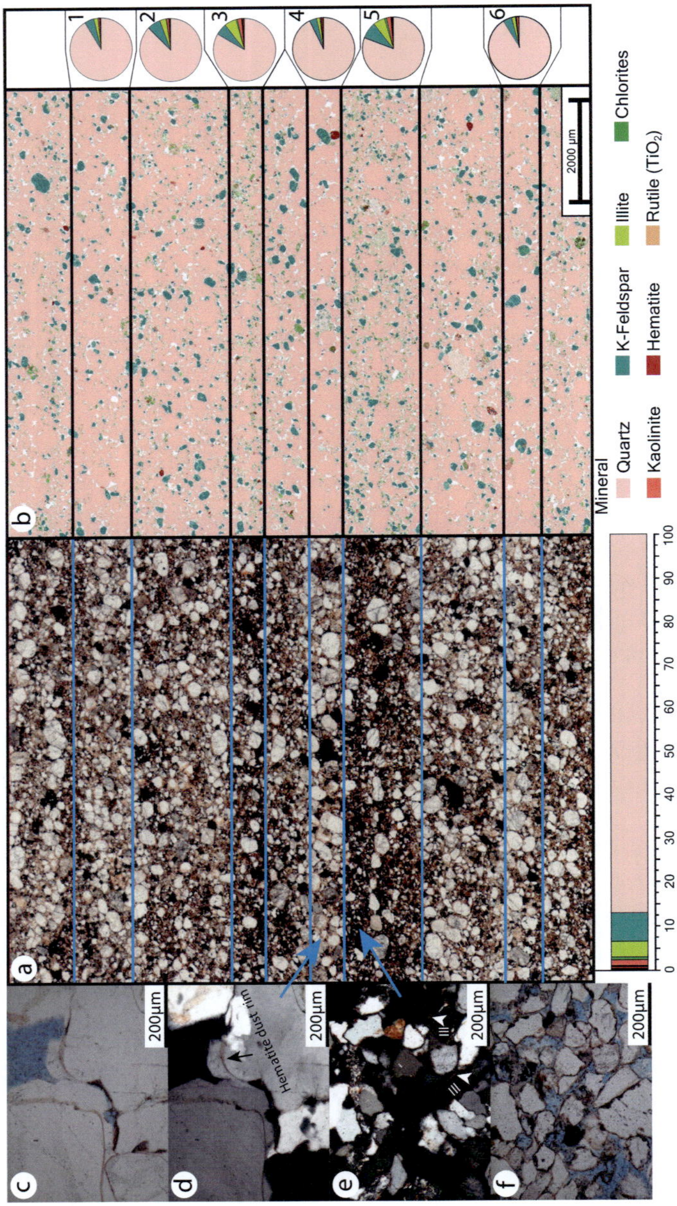

Figure 3.8: Comparison of thick section (100 µm), thin section (30 µm) and QEMSCAN® analyses of sample BQ_2. a + b) The QEMSCAN® analysis showcases the relative enrichment of illite in the finer grained laminae of the dune sample from Bowscar Quarry. The delineation of the textures of illite as a pore filling or grain coating phase was made on thin section images rather than on QEMSCAN® analyses. c, d) Photomicrograph of pigmented hematite dust rims in well cemented, coarser grained fraction of the sample. e, f) Photomicrograph of the better clay coated, finer grained fraction of the pinstripe-laminated sandstone. This part of the sample contains less quartz overgrowth cement. Ill: illite.

Table 3.2: Comparison of QEMSCAN® and point count analyses for sample BQ_2. The pore space from point count analysis has been separately evaluated, to more easily correlate mineralogical information from both analyses. In case of quartz, the pore filling variety represents the quartz cement.

Mineral	QEMSCAN® abundance (%)	Point count abundance (%)	Of which are:		
			pore filling (%)	pore lining (%)	replacive (%)
Quartz	87.04	84.07	12.22		
K-feldspar	6.44	7.04			
Plagioclase feldspar	0.02	0			
Illite	3.54	3.7	0.37	1.48	1.85
Chlorite	0.65	0			
Kaolinite	1.26	1.48	0	0	1.48
Smectite	0.09	0			
Biotite	0.04	0			
Siderite	0.06	0			
Zircon	0.01	0			
Rutile	0.04	0			
Other Heavy Minerals	0.01	0			
Hematite	0.4	1.85	0.74	1.11	
Salt Phases	0.01	0			
Unclassified	0.39	0			
Feldspar-rich volcanic rock fragment	0	0.37			
Shale RF	0	0.37			
Sandstone/Siltstone RF	0	1.11			
Porosity (Additional)	12.86	10			
of which are	Inter granular porosity	8			
	Secondary porosity K-feldspar	2			

3.5.5 Bleaching phenomena around deformation bands and faults

The orientation of deformation bands is illustrated in figure 3.9 for the localities at Lacy's Cave (a), Force Mill (b), and George Gill (c). Two normal fault planes are exposed in Lacy's Cave and their strike matches the orientation of precursor deformation bands (Fig. 3.9 a). In Lacy's Cave and Force Mill, four distinct sets can be distinguished. Conjugate sets of NNW-SSE, NE-SW, and minor W-E strike and NNE-SSW strikes are present in the exposures. The relative amount of deformation bands differs, whereas the exposure at George Gill only features a conjugate NNW-SSE, a minor NNE-SSW, and very minor ENE-WSW striking set. Normal faults and deformation bands, as well as strike slip faults, exhibit the bleaching characteristics (Fig. 3.2 f, g). Exposed strike-slip fault planes at Bowscar and Red Rock Quarry are covered by a white quartzose mineral phase containing slickensides. This evaluation focusses on the two main strike orientations.

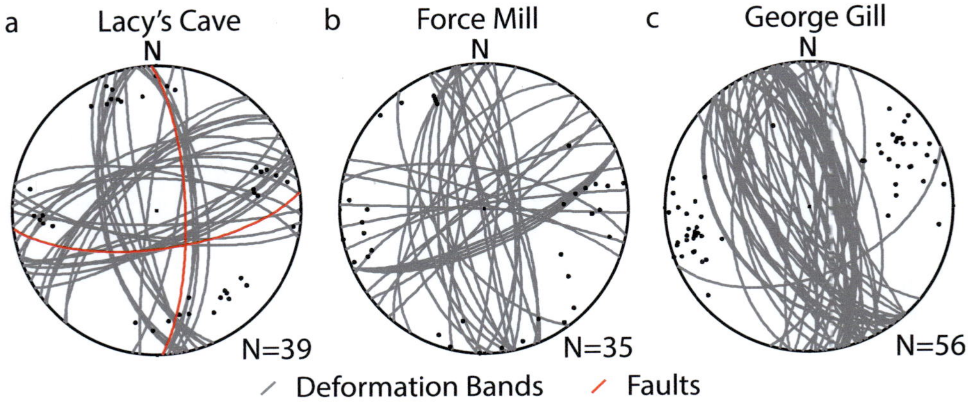

Figure 3.9: Compilation of the orientation of deformation bands from the Vale of Eden. a) The measurements from Lacy's Cave feature three distinct orientations (NNW-SSE, W-E, and (E)NE-(W)SW) and another minor set (NNE-SSW). Two fault planes (red) are exposed and their strike approximately matches the deformation band orientation. b) The deformation band orientation at Force Mill appears more scattered than the nearby Lacy's Cave measurements, but the main orientations are still visible. c) The outcrop at George Gill shows less orientations with a larger set striking NNW-SSE to NW-SE and a very minor set striking ENE-WSW.

The color alteration was observed around normal faults and deformation bands (also called granulation seams in previous reports from this region) in and around Lacy's Cave, including the opposite bank of River Eden (Fig. 3.2 f, 3.10). The host rock some centimeters away from a single deformation band or a cluster of deformation bands is red, whereas occasionally the rock surrounding the deformation band or a cluster and in the vicinity of a normal fault, is greyish white (Fig. 3.10 a, b) or beige (Fig. 3.10 c, d). The beige color alteration was observed around deformation bands striking NE-SW. Deformation bands of all strike orientations, including some of NE-SW strike, appear as thin white seams with a surrounding greyish white halo. Microscopically, two types of deformation bands can be distinguished in samples from Force Mill: One generation exhibits extensive quartz cementation of the cataclastic quartz grains (Fig. 3.11 a-c), whereas the other exhibits no overgrowth cementation on the cataclastic grains (Fig. 3.12 a-d, Tab. 3.3).

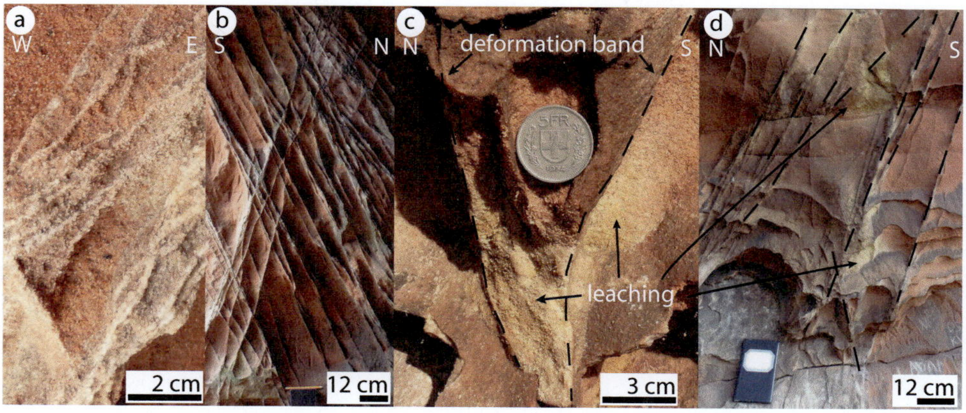

Figure 3.10: Deformation band clusters at Lacy's Cave. a, b) Greyish-white alteration surrounding a deformation band cluster. The host rock in both cases is still red. c, d) Macroscopic image of beige alteration around deformation bands at Lacy's Cave. The alteration is confined to and by deformation bands.

In the sampled greyish-white deformation band striking NNW-SSE (260°/58°), the remains of grain coats are still stained red by hematite and the hematitic dust rims are still present but overgrown by a later quartz cement phase (Fig. 3.11 a-c). The iron minerals present have not been leached in these deformation bands and the color of the alteration is greyish-white (Fig. 3.2 f, 3.10 a, b).

The second sampled generation of deformation bands, striking NE-SW (320°/60°), is often microscopically characterized by the absence of iron minerals on grain surfaces and colorless clay minerals in the vicinity of the deformation band (Fig. 3.12 a-d). Macroscopically, this generation features a beige alteration halo Fig. 3.10 c, d). Some larger grains, including hematite, are still present in the sample. Illite is present in the bleached intervals as a pore filling and grain coating phase (Fig 3.12 a, b, 3.13). Pigmented hematite dust rims very occasionally remain at grain contacts. In the red colored intervals, the illite coatings are visibly stained by hematite (Fig. 3.12 e, f). The cataclastic quartz grains are not affected by quartz cementation, and the deformation bands still appear porous (Fig. 3.12). A model for a paragenetic sequence including this leaching phenomena and faulting is presented in the discussion chapter.

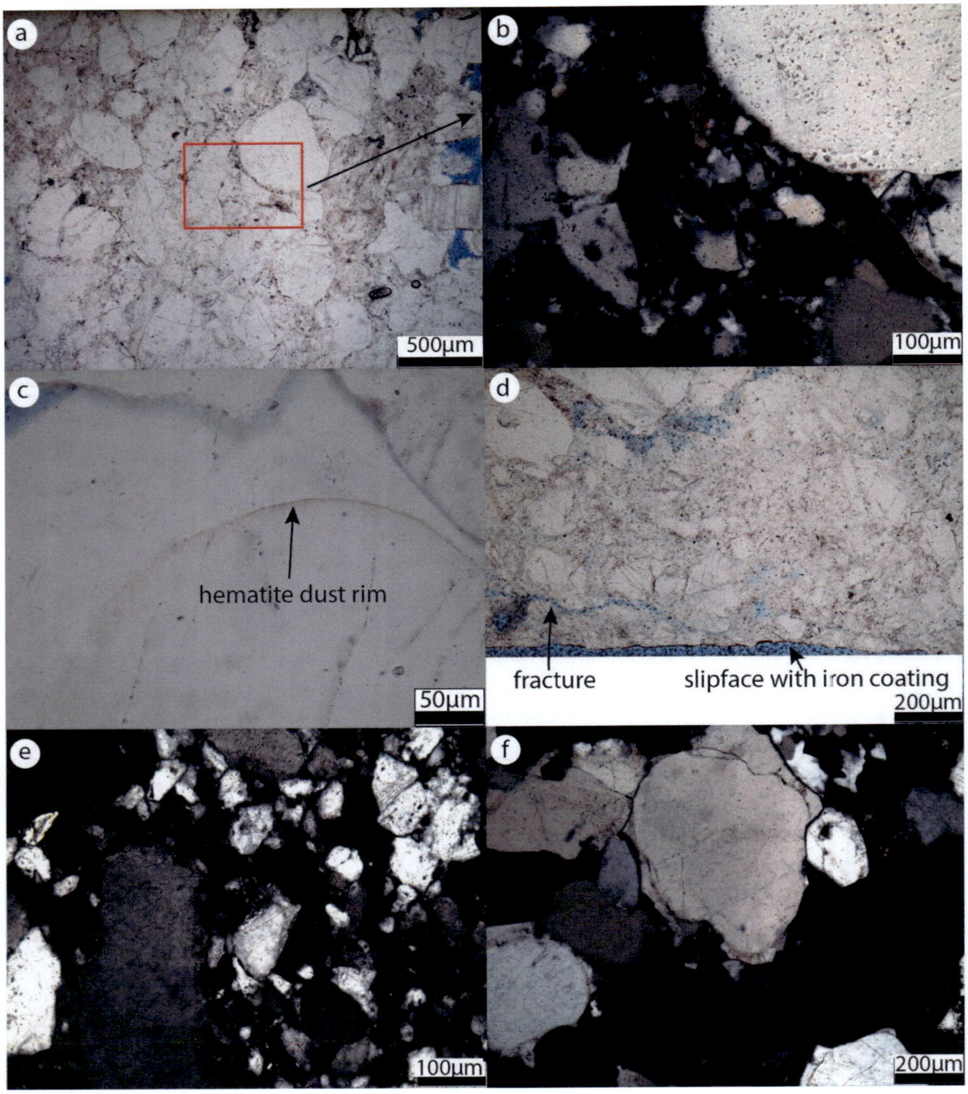

Figure 3.11: Photomicrographs of deformation band from Force Mill and strike slip fault from Red Rock Quarry. a) Quartz cemented deformation band with no retained porosity in sample FM_3. The undisturbed host rock on the right features pores. b) Magnification of quartz cemented cataclastic band (red box in a) showing angular quartz fragments due to cataclasis, which have been quartz cemented (FM_3). c) Directly surrounding the cataclastic band, quartz cement preserves hematite dust rims on detrital grains (FM_3). d) Overview of cataclastic slipface from strike slip fault in Red Rock Quarry (see Fig. 3.2 g, RR_2). Some porosity is retained including fracture porosity. e) Magnification of cataclastic quartz grains featuring quartz overgrowth cementation in sample RR_2. f) Undisturbed host rock around slipface, being quartz cemented and featuring hematite dust rims around detrital grains in sample RR_2.

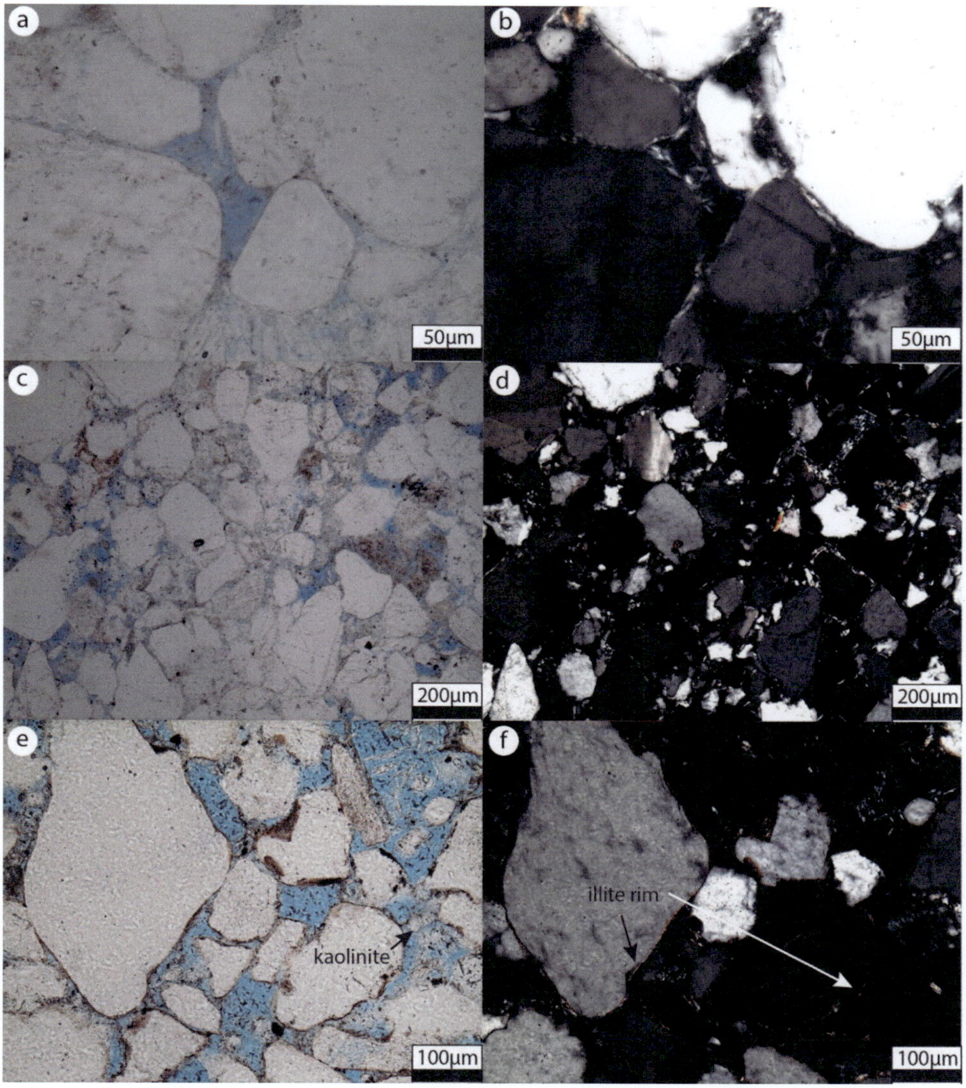

Figure 3.12: Cataclastic deformation band from Force Mill (all images from sample FM_Alt) a, b) Illite coatings from a bleached part of the lithology close to a deformation band at Force Mill. The illite coatings can clearly be distinguished by the interference colors. c, d) Overview of the cataclastic deformation band. Some porosity is retained within the band. Additionally, the thin hematite dust rims are absent. e, f) Approximately two mm away from the deformation band cluster, the illite rims still appear with a reddish-brown stain by hematite. Pore-filling kaolinite is present as an alteration product of feldspars and appears unillitized in this case.

Table 3.3: Comparison of two deformation band generations based on macroscopic and microscopic alterations. Samples from both major strike directions appear grey to white, whereas the beige alteration was confined to the NE-SW striking set.

Sample	Fig. 3.10 a+b, Fig. 3.11 a-c	Fig. 3.10 c+d, Fig. 3.12 a-d
Color of halo	Greyish white	Beige
Strike	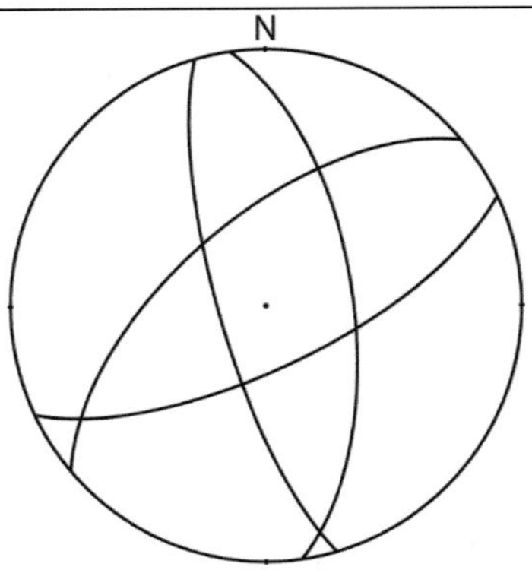	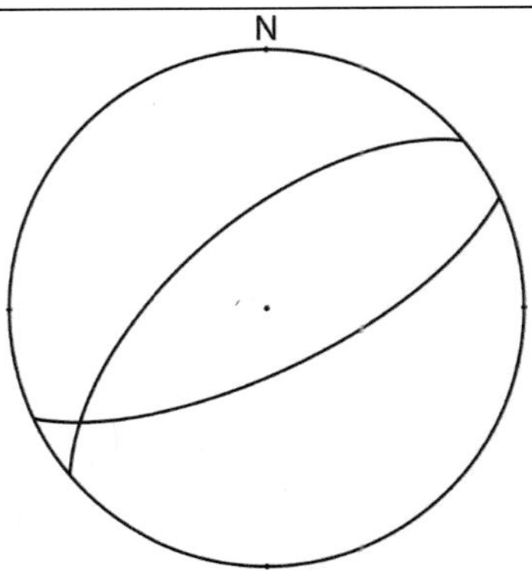
Cataclasis	Yes	Yes
Quartz cement	Yes	No
Bleaching	No	Yes
Hematite dust rims	Yes (inside cataclastic deformation band)	No
Timing	Before burial diagenesis	After burial diagenesis

Mineralogically, the clay coatings in both the bleached and red parts of the lithology are identified as illite by QEMSCAN® analyses. Microscopically, the red color is created by small hematite agglomerates on the grain surfaces. The QEMSCAN® analyses also highlight a relative increase in illite and kaolinite abundance in the beige alteration halos as compared to the red sandstones (Fig. 3.13). The illite abundance is especially increased in the attrition material inside the deformation band (bottom Fig. 3.13 b). In the unbleached part of the formation, the deformation band is mainly characterized by broken K-feldspar grains (blue-green grains at the bottom in Fig. 3.13 a). Additionally, a difference in grainsizes can be observed. The unbleached interval features a slightly larger average grainsize of 0.20 mm compared to 0.18 mm in the bleached interval. Additionally, the sorting after Folk & Ward (1957) in the unbleached part is slightly poorer at 0.72 as compared to 0.74 in the bleached part, still being moderately sorted.

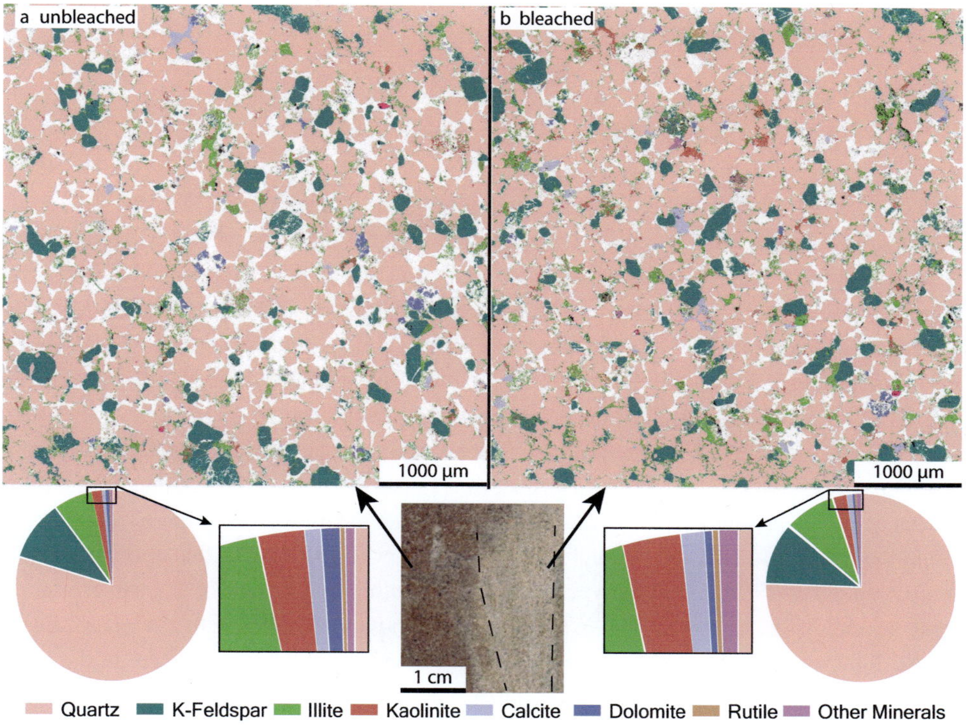

Figure 3.13: Comparison of the mineralogical suite of regular (a) and bleached (b) sandstone around deformation bands at a beige alteration zone in the Penrith Sandstone (center image) using QEMSCAN® in Sample FM_Alt. The dashed line in the specimen photo outlines the multistrand cataclastic band. The bleached interval (b) contains relatively more illite (green) and kaolinite (red) than the unbleached rock (a). The white area in between grains represents porosity filled by an epoxy resin.

Small strike slip faults and clusters of strike slip faults with an alteration zone and an unknown offset have been observed in Red Rock and Bowscar Quarry. Exposed NE-SW, NNW-SSE, and WSW-ENE striking fault planes exhibit slickensides on quartz mineralization which point towards both dextral and sinistral motion. Occasionally, a reddish-brown coating can be identified on the fault planes including slickensides. The fault sample from Red Rock Quarry exhibits a max. 0.5 mm wide cataclastic zone at the sinistral strike slip fault plane (Fig. 3.11 d, e, f). The relatively undisturbed host rock contains small amounts of illite and hematite on the grain surfaces. The cataclastic zones are characterized by a visible decrease in grainsize (Fig. 3.11 e, f) but hematite dust rims covering quartz grain surfaces are still present below syntaxial quartz overgrowths. Syntaxial quartz cement growth is pervasive in the cataclastic zones (Fig. 3.11 e). However, some porosity is still retained within the cataclastic zone, especially in fractures.

3.6 Discussion

3.6.1 Petrography and QEMSCAN®

The correlation of standard petrographic point counting analysis and QEMSCAN® analyses is possible, but only to a limited extent. The application of QEMSCAN® analyses in a petrological workflow of sedimentary rocks has been previously established (e.g. Armitage et al., 2013). However, multiphase grains, i.e. rock fragments, cannot be easily identified as one grain. Replacive phases of clay minerals or carbonate cements cannot be easily distinguished as such. The detection of multi-mineralic rock fragments, which can act as cation donors during diagenesis and bear information on the source area, is not easy made. This is relevant for interpreting the paragenetic sequence and provenance of the sedimentary rocks. Additionally, the differentiation of mono- and polycrystalline or undulose grains cannot be distinguished by QEMSCAN® analyses although they have an impact on syntaxial quartz cementation (Waugh, 1970a; Lander et al., 2008). The detection of mineral phases does not differentiate between detrital and authigenic phases of the same mineral type, an analysis easily performed in case of quartz on these samples with a standard, transmitted light microscope or conventional or SEM-based cathodoluminescence. In the evaluation of clay coat coverage, the current technology is not able to automatically and sufficiently detect grain coating illite in polished thin section samples. Occasionally, the coat cannot be detected due to an interference of the grain coating phase and the substrate grain in EDX-analyses. This is due to the small size of particles and the relative size of the excitation volume during EDX-analyses (e.g. El-Gomati et al., 2011). Markedly, the presence of opaque iron phases (hematite) is overrepresented by point count analyses, due to the presence on all grain surfaces as pigmented hematite rims. Since these pigmented rims are mainly overgrown by a syntaxial quart cement they are rarely detected in QEMSCAN® analyses. Some minerals with low concentrations are well visualized by the QEMSCAN® analyses (e.g. chlorite). Including the assessment of secondary porosity restricted to certain mineral phases, the standard petrographic analyses are superior to QEMSCAN® analyses.

3.6.2 Paragenetic sequence

Integrating the petrographic and lithological observations in correlation with literature results in the following paragenetic sequence (Fig. 3.14).

3.6.2.1 Early diagenesis

Earliest diagenetic alteration was the formation of hematite and clay rims as observed for other red beds and eolianites (van Houten, 1973; Ajdukiewicz et al., 2010), with considerable impact on the inhibition of syntaxial quartz cement growth in the studied region (Fig. 3.7). Climatic changes might have had an impact on the presence and extent of clay coats in the Penrith Sandstone. Mader and Yardley (1985) postulated a shift to more humid/semi-arid conditions during the late Permian in this basin. This shift is marked by an increase in the periodicity of ephemeral fluvial intercalations in the Rotliegend lithology towards the Zechstein Eden Shales (Mader and Yardley, 1985). This model of the climatic evolution towards less arid conditions would support the increase in grain coating illite now present in the formation (see table 3.1). Waugh (1970b) already reported a less silicified sandstone in the region, where eolian sandstones are associated with waterlain deposits in the studied area. Another factor impacting the varying amounts of grain coat coverage of illite could be the migration activity of a dune. Ajdukiewicz et al. (2010) proposes an abrasion of coating materials during dune migration, where only smaller grains are able to retain their coating (Wilson, 1992). Pigmented hematite and minor amounts of clay minerals can be preserved in small surface roughnesses during the transport (Wilson, 1992; Fischer et al., 2013). This could explain the observed differences in samples from Red Rock and Bowscar Quarry (Fig. 3.8). Additionally, the capillary retention capacity of finer grained layers can explain such differences if the clay minerals are infiltrated (Smith et al., 1930). Here, the retained water allows the accumulation of infiltrated clay minerals in an early diagenetic environment, whereas the fluids percolate more easily through the matrix created by coarser grains.

Hematite, responsible for the red color of the lithology, is a proxy for arid to semi-arid conditions during eolian deposition (van Houten, 1973).

An early quartz cementation phase was hypothesized by Waugh (1970b) and linked to the habit of observed quartz overgrowths (open box in quartz cement paragenesis). Based on observations and theories by Waugh (1970b), Turner et al. (1995) links the observed growth patterns in cathodoluminescence of samples from the Vale of Eden to two discrete quartz cementation phases, whereas the images and descriptions resemble non-euhedral and euhedral structures experimentally created by Lander et al. (2008, their Fig. 9) in a continually grown cement. The early quartz cement generation might thus be disputed based on their arguments.

Where carbonate cements, here calcite, dolomite, and siderite, are still present, they appear pore filling and framework stabilizing and are interpreted to be of an early diagenetic origin. Due to the arid depositional environment during the Rotliegend and limited textural evidence, the carbonate formation is thought to occur during early diagenesis and contemporaneously to deposition (Morad et al., 2000). Small rhombic pores inside quartz cements were previously linked to early dolomite cementation (see Fig. 8 in Waugh, 1970b; Turner et al., 1995). An early anhydrite phase is evident from textural relations appearing as framework stabilizing pore fillings.

Feldspar dissolution and kaolinite authigenesis is interpreted to occur in an early and a late stage of diagenesis (Lanson et al., 2002). The early phase is marked by later burial diagenetic alteration to illite.

3.6.2.2 Burial diagenesis

Quartz overgrowth generation during burial is responsible for quartz authigenesis in both the host rock and previously faulted rocks. Quartz overgrowth is locally inhibited by the presence of extensive grain coat coverages. Variations in quartz cementation can be explained by the localized occurrence of illitic coatings (see Fig. 3.7), which are described to limit syntaxial overgrowth cementation (Storvoll et al., 2002). Earlier work by Waugh (1970b, a) already qualitatively described the presence of illite and hematite on grain surfaces but lacked the quantitative correlation (Fig. 3.7). The hematite stain is present at all exposures but the occurrence of illite and its grain coating coverage varies strongly within the sampled parts of the Rotliegend section, but seemingly increases towards the Zechstein deposits.

The fluid chemistry during quartz cementation is rather well constrained by fluid inclusion analyses of samples from the Penrith Sandstone Formation by Turner et al. (1995) and points towards an alkaline milieu, reaching maximum temperatures of approximately 120 °C. The alteration of precursor clay phases on grain surfaces towards illite and the illitization of feldspars is also interpreted to occur in this higher temperature domain (Worden and Morad, 2002). The replacement of parts of the first kaolinite phase by illite is kinetically controlled during burial diagenesis (e.g. Lander and Bonnell, 2010). This alteration is additionally controlled by the amounts of K-feldspar still present during burial diagenesis, acting as a source for potassium (Ehrenberg and Nadeau, 1989). Since all three reactants are available in the samples, the process is not expected to have ceased during burial diagenesis.

Evident from magnetic remanence experiments on samples from the Penrith Sandstone Formation by Turner et al. (1995), hematite authigenesis occurred continuously throughout the burial history.

The alteration of rock fragments and pore filling precursor clay minerals to chlorite is expected to be a burial related alteration (Boles and Franks, 1979; De Ros et al., 1994; Hillier, 1994; Remy, 1994; Worden and Morad, 2002). The amounts of chlorite and the scattered distribution within the sample question its importance for the overall paragenetic sequence.

3.6.2.3 Uplift diagenesis

Remnants of carbonate and sulfate cements in these Rotliegend rocks from exposures appear partly dissolved as a result of interaction with meteoric waters. A second, unillitized phase of kaolinite is interpreted to have formed during late uplift, due to exposure related diagenesis and interaction with meteoric waters (Lanson et al., 2002).

3.6.2.4 Structural development

The general timing of fault movement is unclear (Fowles and Burley, 1994). However, the timing of the large scale Pennine Fault System and Dent Fault System is well constrained (Underhill et al., 1988; Turner et al., 1995). Relative dating of some structural elements is possible around the Armathwaite-Cleveland Dyke. The Tertiary Armathwaite-Cleveland Dyke (Emplacement 58.4 ± 1.1 Ma, MacDonald et al., 1988) is also offset by NE-SW to ENE-WSW striking normal faults (Fowles and Burley, 1994). The similar strike of deformation bands and larger scale faults indicates that they were formed contemporaneously triggered by the same geotectonic event and within a similar stressfield (e.g. Antonellini and Aydin, 1995; Johansen and Fossen, 2008; Fossen and Rotevatn, 2012).

Only two generations of deformation bands can be linked to normal faulting occurring at a specific time. The NNW-SSE striking set (Fig. 3.9 a-c) most likely resembles a phase of formation during early subsidence controlled by the large Pennine Fault system (Underhill et al., 1988). This fault was active during the deposition of the Penrith Sandstone, and during the Tertiary (Arthurton and Wadge, 1981). Initiated during the Alpine orogeny, the down-west faulting reactivated many older faults (Arthurton and Wadge, 1981). The scatter of measured orientations at the George Gill site was related to the distance of deformation bands to the normal fault (Guo et al., 2009).

The NE-SW striking set (Fig. 3.9 a-c), which appears partly uncemented, can be related to a late phase of normal faulting also displacing the Armathwaite dyke further north (Fowles and Burley, 1994). Additionally, the NE-SW striking faults mostly abut against NNW-SSE striking faults (Fowles and Burley, 1994). This abutting relation supports a younger age for the NE-SW striking faults (e.g. Nixon et al., 2014).

Both major faulting orientations present in the Vale of Eden are thought to be inherited from older strata (Arthurton and Wadge, 1981). The early activity of the NNW-SSE striking fault orientation is supported by macroscopic similarities with the NNW-SSE striking deformation bands (see table 3.3). Fowles and Burley (1994) point out the similar strike of the Dent Fault zone in the south, whereas Arthurton and Wadge (1981) underline the importance of basal normal faults separating the depositional environments of the Carlisle- and Vale of Eden Basins north of the area.

The minor E-W striking set of deformation bands (Fig. 3.9 a, b) is also reflected by a W-E striking normal fault mapped by Fowles and Burley (1994). However, the timing is not further constrained.

The sampled quartz cemented, unbleached, potentially earlier set of deformation bands approximately matches the strike of the large Pennine Fault. The sampled uncemented, bleached, and relatively porous, potentially later deformation band approximately match the strike of normal faults displacing the Tertiary dyke.

The observed small scale strike slip faults from Bowscar and Red Rock Quarry could not be correlated to larger scale faults in the region. However, the presence of quartz slickensides favors a development during burial diagenesis. The timing of strike slip faults is inferred to occur at a time during burial diagenesis since quartz cementation in cataclastic zones is present but not as pervasive as in completely cemented bands (Fig. 3.11 a-c). Additionally, small amounts of hematite are still retained underneath quartz cements (Fig. 3.11 d-f), thus they have to predate the leaching process discussed below. This allows a change of the geotectonic stressfield from normal to strike slip faulting. The observed local color alteration (Fig. 3.2 g) is interpreted to be due to quartz cementation of cataclastic grains, similar to the greyish-white alteration surrounding the deformation bands.

3.6.3 Bleaching phenomena around deformation bands and faults

3.6.3.1 Greyish-white alteration

The observed greyish-white color alteration of a NNW-SSE striking deformation band cluster at Force Mill and Lacy's Cave is interpreted to be mainly controlled by a completely pore-filling quartz cement phase which preserves hematite dust rims (see Fig. 3.11 a-c). This phase of deformation band formation is thus not affected by bleaching processes. They predate syntaxial quartz cementation during burial diagenesis and extensional tectonics and subsidence in the region. This color alteration was also observed around some deformation bands striking NW-SE.

3.6.3.2 Beige alteration

The observed beige alteration halo around NE-SW striking deformation bands is interpreted to be a product of hematite dissolution and transport, since little to no hematite is present on grain surfaces or intermixed with the illite coatings (Fig. 3.12 a-d).

The observed leaching of the host rock surrounding deformation bands does not match well with the published hypothesis of which fluids impacted the formation. Turner et al. (1995)

present an argument for alkaline fluids originating from the Eden Shales Formation in the hanging wall of the formation. These fluids are reported to be able to carry large saturations of silica in solution, and the salinity of fluid inclusions within quartz overgrowths supports this model (Turner et al., 1995). However, alkaline fluids tend not to dissolve hematite under oxidizing conditions which are discussed to have been prevailing during quartz cementation (Turner et al., 1995; Takeno, 2005). As presented by Parry et al. (2004), reducing organic fluids and hydrocarbons are able to result in sandstone bleaching of red beds. More generally, the dissolution of iron-bearing minerals is favored by acidic conditions and reducing fluids (Tucker, 2001; Takeno, 2005).

The different fluid chemistries need to accommodate 1) SiO_2–rich, alkaline, oxidizing, and saline conditions during quartz cementation and 2) acidic and reducing conditions during bleaching.

A model could be the temperature controlled release of acidic fluids during breakdown of kerogen from e.g. Carboniferous coal seams at temperatures of approximately 100°C (cf. Platt, 1993). The Appleby Group in the northern part of the studied area unconformably overlies Upper Carboniferous strata containing the Coal Measures and Millstone Grit including coal seams (see Arthurton and Wadge, 1981; Hughes, 2003). The migration of these fluids into the Permian strata and the formation of the deformation bands and faults has to postdate extensive quartz cementation, affecting the freshly broken grains of the then and still uncemented cataclastic deformation bands. Some examples in this study and from Fowles and Burley (1994) exhibit no syntaxial overgrowth cementation inside cataclastic deformation bands, which seem to represent a later generation of cataclastic deformation bands. In this study the leaching was examined on a deformation band sample striking NE-SW. Normal faults of a similar strike displacing the Armathwaite dyke point towards a formation postdating the emplacement during the Tertiary (Fowles and Burley, 1994). This would require a phase of extensive tectonics featuring normal faulting briefly interrupting the uplift of the area since the Tertiary.

Considering the timing of fluid flow, Fowles and Burley (1994) assume the deformation bands and fault clusters to only be transmissive during and shortly after the process of faulting. Their results however show a zone of increased permeability around deformation bands still present at outcrops. The leaching in the closely fault-associated host rock should nevertheless be contemporaneous with or slightly postdating the formation of the bands. However, the absence of both quartz overgrowth cement and clay coatings on freshly broken quartz surfaces at the same time suggests formation at temperatures below approximately 70 °C and postdating pervasive quartz cementation.

Since both the red and beige sandstone within sample FM_Alt exhibit illite coatings (Fig. 3.13), the influence of migrating fluids on these coatings is very minor and only results in a discoloration. The effectivity on their inhibition potential on syntaxial quartz cement seem not affected. However, this evaluation is limited since the leaching by fluids postdates the quartz cementation phase.

The limited lateral extent of the beige alteration could be explained by pervasive carbonate and sulfate cementation of the strata in the subsurface still present in shallow boreholes (Arthurton and Wadge, 1981; Turner et al., 1995). The deformation bands only enhanced fluid flow in a limited zone parallel to its extent (Fowles and Burley, 1994).

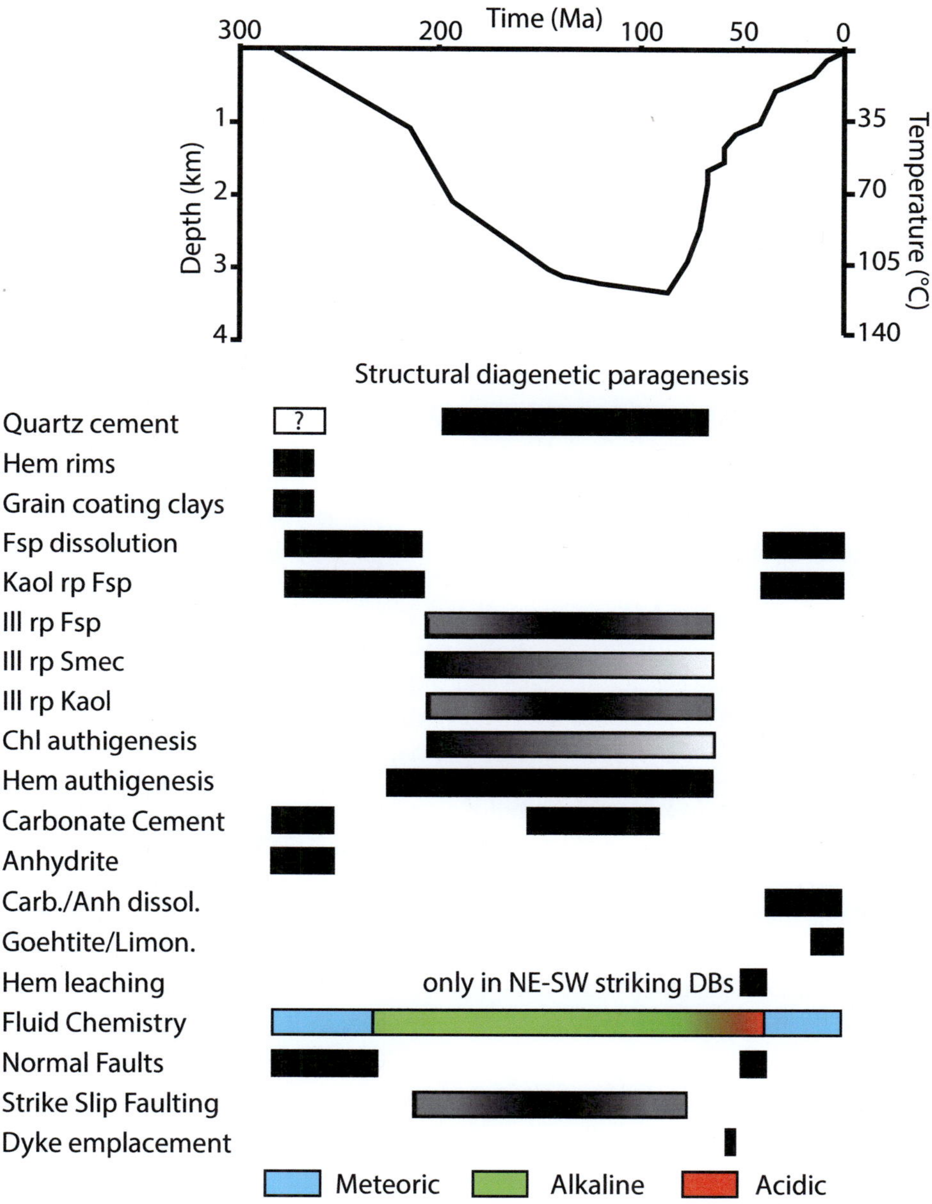

Figure 3.14: Reconstructed burial curve from Turner et al. (1995) including the presented paragenetic sequence. Replacements are indicated by the term rp, with the replacing phase being named first. The discussed early phase of quartz cement is presented with a question mark. The potential illitization of smectitic precursor grain coats is interpreted to stop once the precursor smectite is consumed, as indicated by the shaded box. Transformation to chlorite is also expected to cease once the reactants are consumed. The possible formation of strike slip faults is indicated as a range during quartz cementation. Hem - hematite; Fsp - feldspar; Kaol - kaolinite; Ill - illite; Smec - smectite; Chl - chlorite; Carb - carbonate; Anh - anhydrite.

3.7 Conclusions

This study has shown that grain coat coverages in the studied samples correlate well with observed quartz cement volumes and highlight the importance of clay coatings for the evaluation of controlling factors during sandstone cementation. They are a key to decipher the structural and diagenetic development as they retained hematite within their fabric and inhibited quartz cement. Only where the illite – hematite mixture is in contact with the pore fluid, it can record the localized bleaching postdating pervasive quartz cementation.

At least two phases of normal faulting and associated deformation band formation occurred and can be dated relative to burial diagenetic alterations. The first, unbleached, NNW-SSE and NE-SW striking orientation predates or syndates early burial diagenesis. The second phase resulted in a bleached (beige), NE-SW orientation postdating burial diagenetic alterations and includes evidence of hematite leaching during fault focused fluid migration. The color alteration and retained microporosity of strike slip cataclastic fault rocks, while featuring quartz cements, indicates formation during burial diagenesis.

Linking the structural and diagenetic evolution is important for hydrocarbon migration and entrapment. The suitability of the presented exposures as direct analogs for subsurface reservoirs is reduced since uncemented cataclastic deformation bands are not expected in reservoirs at depth with temperatures above approximately 70 °C. Their character as paleo fluid pathways however, is still indicated by leaching around faults and deformation bands.

QEMSCAN® can deliver valuable information on low-concentration detrital and authigenic phases in sandstones but cannot significantly improve the assessment of factors controlling reservoir quality like grain coatings which are at the lower limit of resolution for this method. The presence of mono- or polycrystalline quartz grains can yet only be differentiated by methods that can visualize crystal optics. A combination of transmitted light microscopy and/or cathodoluminescence with QEMSCAN® could however, result in an improvement of quantitative and qualitative analysis during the assessment of diagenetic alterations in sandstones.

4 Distribution of grain coatings in fluvio-eolian red beds, case study Rotliegend, Southern Permian Basin

4.1 Abstract

Reservoir quality in deeply buried sandstones is strongly affected by grain coatings. To predict and infer the quality of undrilled reservoirs, constraints are needed to predict their occurrence. The objective of the study was to derive dependencies of the formation of grain coatings which are the basis for a universal predictive tool, i.e. applicable to the entirety of studied samples. Like other early diagenetic or even syn-depositional alterations, their presence and characteristics should be affected by the depositional environment and processes controlled thereby. The depositional environments, grain size, sorting, and the impact of volcanic rock fragments only locally correlate with the amount of grain coat coverage. On a regional to basin-scale (Southern Permian Basin and an adjacent sub-basin) variations emerge. Samples originate from northern Germany, the Netherlands, and the United Kingdom. This diminishes the availability of a universal predictive tool to assess reservoir quality based on sedimentary properties for the siliciclastic, fluvio-eolian Rotliegend in the Southern Permian Basin. The use of universal dependencies results in deviations during reservoir quality modeling and prediction using Touchstone™. To prevent these deviations, it is imperative that the reservoir quality is estimated based on nearby analog wells in this Rotliegend setting.

4.2 Introduction

Grain coatings exert an impact on the formation of syntaxial quartz overgrowth cements and are thus a primary control on reservoir quality in siliciclastic reservoirs (Storvoll et al.,

2002; Ajdukiewicz and Larese, 2012). However, the main control is not the simple presence of grain coatings, but how much of the surface area of substrate grains is covered (Heald and Larese, 1974; Walderhaug, 1996).

The formation of clay coatings on quartz grains is mainly governed by the supply of (1) clay minerals to act as precursory mineral phases on grains, (2) the influence of meteoric and formation waters, and (3) the subsequent diagenetic pathway (Ajdukiewicz et al., 2010). Precursory clay coatings are discussed to be formed either by illuviation or infiltration of clay, blown into the sedimentary system and percolated into the sediment by water (Seemann, 1982; Ajdukiewicz et al., 2010). Alternatively, they can form in-situ during late eo- to mesodiagenesis as a result of dissolution of unstable alumosilicates or rock fragments like volcanic rock fragments and precipitation on surrounding detrital grains (Anjos et al., 2000; Liewig and Clauer, 2000; Grigsby, 2001; Anjos et al., 2003; Bahlis and De Ros, 2013). Clay coatings at depth or of exhumed lithologies are most likely a product of recrystallization of precursory clay phases (Aagaard et al., 2000). This recrystallization is described to be responsible for the final observed coat coverage (Aagaard et al., 2000). Overall, differences in the depositional environment could exert a control on the presence of precursory clay minerals, also represented in samples which underwent burial. Additionally, grain size, sorting, and detrital grains may have a direct impact on the presence and extent of grain coatings which is then altered during diagenesis (Walker, 1979; De Ros et al., 1994; Remy, 1994; Aagaard et al., 2000; Bloch et al., 2002; Anjos et al., 2003; Ajdukiewicz et al., 2010; Darbha et al., 2012; Bahlis and De Ros, 2013).

In some cases, the occurrence and the type of clay coats can be associated with the depositional environment (Pittmann et al., 1992; Dowey et al., 2012). Some studies focused on other controlling factors in eolian systems and derived the influence of meteoric waters during illuviation (Moraes and De Ros, 1992) and the role of ground water chemistry (Ajdukiewicz et al., 2010). However, the influence of depositional environments and the accompanying changes concerning grain coatings within the eolian system have never been systematically studied on a large scale to derive the controlling factors.

Differences in early diagenetic assemblages governed by the depositional environment including but not limited to eolian systems are well studied and described for multiple settings (Seemann, 1982; Anjos et al., 2000; Morad et al., 2000; Ramm, 2000; Bloch et al., 2002; Ketzer et al., 2002; Elias et al., 2004; El-Ghali et al., 2006; Hammer et al., 2010; Morad et al., 2010; Bjørlykke, 2014; Henares et al., 2014; Khalifa and Morad, 2015; Marchand et al., 2015; Lan et al., 2016). However, for grain coatings, their distribution and occurrence is more uncertain (Haile et al., 2015). Although grain coatings are known as major inhibitors for quartz cementation in the pore space (Heald and Larese, 1974; Ajdukiewicz and Larese, 2012), distributions for only few depositional environments could be established (Ehrenberg, 1993).

Sorting and grain size strongly affect the porosity of sediments (Rogers and Head, 1961). Both are evaluated concerning their impact on the presence of clay coatings and their coverage. In regional studies, a larger grain coating coverage has been linked to smaller grain size (Bukar, 2013, and references therein). Additionally, the positive impact of rock fragment dissolution on the authigenesis of chlorite coats is evaluated (e.g. Anjos et al., 2003).

Overall, if the depositional environment, grain size, sorting, detrital grains, and the diagenetic overprint have a direct impact on the presence and extent of grain coatings, they should be valid for all the studied settings.

The delineation of universal trends linking coat coverage with those parameters within the fluvio-eolian system could prove beneficial to the reservoir quality assessment and prediction modeling. The term universal in this case reflects the ability to sufficiently reproduce observed and measured amounts. To provide a database to delineate possible controls affecting the distribution and extent of clay coats, we studied Permian Rotliegend red bed deposits of four core areas, both from surface and subsurface samples of the Southern Permian Basin and an adjacent basin (Fig. 4.1).

We establish a database to highlight the predictive capabilities of universal trends of coat coverage versus depositional setting, grain size, sorting, detrital grains, and include the diagenetic overprint. Calibration plots from the Touchstone™ modeling software are prepared to compare the modeling results of measured data from a selected complex reservoir with universal grain coat coverages.

4.3 Geological setting

The deposits of four different study areas of fluvio-eolian Permian Rotliegend strata across central Europe representing the Permian Basin System and associated basins have been comparatively studied. Samples from the surface and subsurface originate from Germany, Dutch North Sea, and the United Kingdom. The presence of grain coatings was established in all of the settings. The studied areas (Fig. 4.1) are:

(1) five wells penetrating the Rotliegend Bahnsen Sandstone Member northeast of Hannover in central northern Germany,
(2) five wells penetrating the Rotliegend Slochteren Sandstone in two adjacent blocks (K10 and K13) offshore the Netherlands,
(3) one outcrop of the Rotliegend Parchim Formation in Bebertal (Schwentesius Quarry), northwest of Magdeburg, eastern Germany, and

(4) five outcrops of the Rotliegend Penrith Sandstone Formation around Penrith, Cumbria, N-England in the United Kingdom (Bowscar Quarry, Red Rock Quarry, Quarry at Salter Hill, Force Mill and Lacy's Cave).

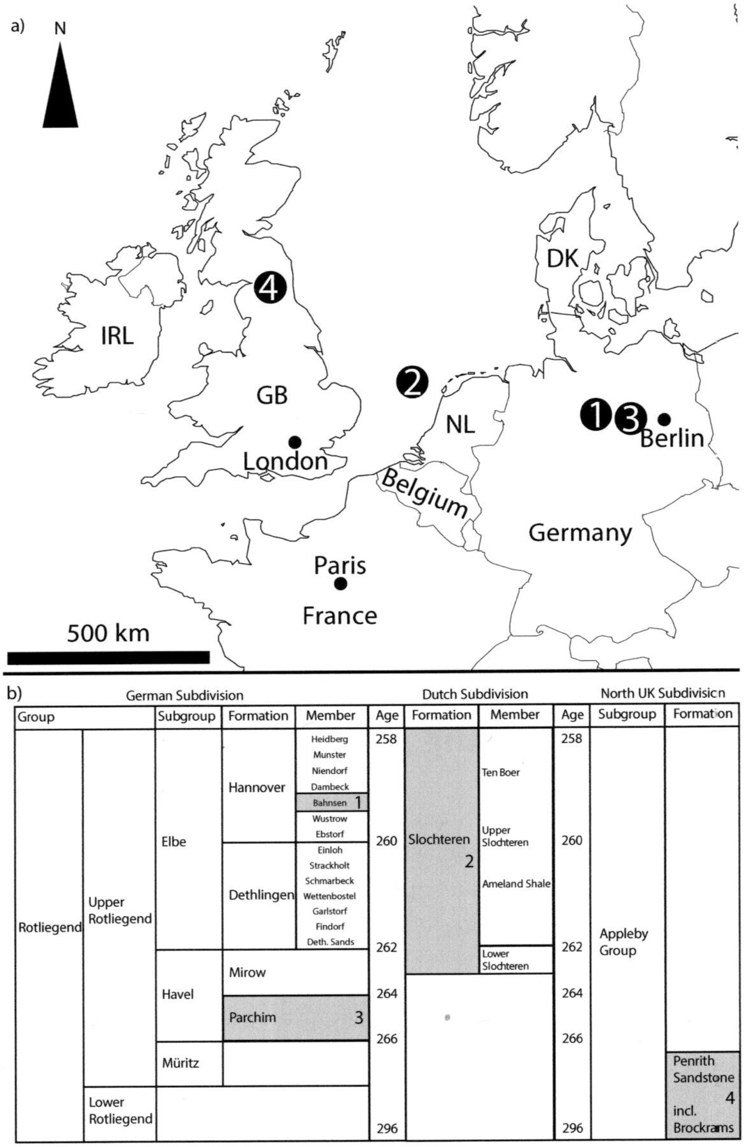

Figure 4.1: a) Location of the four studied areas in Europe. (1) wells NE of Hannover, (2) Dutch offshore blocks, (3) Schwentesius Quarry in Bebertal, (4) Penrith Sandstone Formation in N-England. b) Stratigraphic ages of all studied deposits (grey boxes) (based on (Menning, 1995; Deutsche Stratigraphische Kommission, 2002; BGS, 2016)). Deth. Sands: Dethlingen Sands.

(1) NE-Hannover sample set (wells A, B, C, D, E)

The sediments of the Bahnsen Sandstone were deposited in a semi-arid eolian setting comprising of eolian dune, sandflat and lacustrine deposits (Antrett, 2011). The overall setting is additionally controlled by cyclic lacustrine intercalations as studied in other Rotliegend settings (Rieke, 2001; Gast et al., 2010). Based on a classification scheme, described in the methods section, the cored lithotypes were interpreted as eolian dune, dry, damp and wet sandflat or sandflats, lake margin, low-energy fluvial deposits and lacustrine deposits. The Rotliegend of the area is overlain by Zechstein evaporites, occasionally forming salt pillows. Wells A, B, C, D and E are situated on the Pompeckj Block which is part of the Permian Basin System formed as a response to early Permian rifting (Gast et al., 2010). The samples are presently at 4264 to 4570 m TVD.

(2) Dutch Slochteren Sandstone samples (wells F, G, H, I, J)

The sediments of the Slochteren Formation were deposited in the southwestern part of the Southern Permian Basin. This area was inverted during the late Jurassic and early Cretaceous and now forms a transition between the Broad Fourteens Basin and the Cleaver Bank High (Verweij, 2003). Depositional environments are similar to those studied in Germany (NE-Hannover and Bebertal. The subsequent development varied slightly for the different wells F-J. Four wells (F-I) are situated in block K10 and one well is situated in block K13 (well J). The burial and subsequent erosion differs from a well at the margin of the Broad Fourteens Basin (well J) as opposed to the deposits of the Cleaver Bank High (Verweij, 2003, Fig. 18). The additional late Cretaceous inversion of parts of the Broad Fourteens Basin and salt movement is especially present in well J (Roos and Smits, 1983; Verweij, 2003). Samples are currently between 2479-3117m TVD. They are currently at maximum burial depth. Generally, the depositional system was controlled by wind and meteoric water supply in this arid to semi-arid setting (Glennie, 1983).

(3) Bebertal outcrop analog (Schwentesius Quarry)

The sediments present in an active quarry near Bebertal represent fluvio-eolian deposits from the Rotliegend Parchim Formation. The deposits in Schwentesius quarry comprise of alluvial deposits, fluvial sheetsands, sandflat, lacustrine, and eolian dune deposits (Fischer et al., 2007). Initial fast and deep burial was followed by exhumation of the Flechtingen Block during the Palaeogene (Fischer et al., 2012). Maximum temperatures as estimated from Zircon Fission track dating exceed 200 °C (Fischer et al., 2012). The studied lithologies develop from alluvial and fluvial deposits at the base into eolian dune sandstones with few lacustrine intercalations (Gaitzsch et al., 2004).

(4) Penrith Sandstone Formation, N-England

The Penrith Sandstone Formation was deposited in the Permian Vale of Eden halfgraben, unconformably overlying Upper Carboniferous Coal Measures and Stainmore Formation sandstones and Lower to Upper Carboniferous limestones of the Alston Formation and Great Limestone Member (Macchi, 1981). Deposits from the Cisuralian comprise alluvial and fluvial deposits and eolian dune deposits. The deposits are intercalated by sandflat and fluvial deposits (Macchi, 1981) which were not sampled in this study. Towards the Zechstein, deposition was periodically impacted by fluvial deposition, triggered by an increase in meteoric water supply, leading to the deposition of ephemeral fluvial channel sediments (Mader and Yardley, 1985). The maximum burial reaching a depth of approximately 3.5 km was succeeded by uplift from the Upper Cretaceous onwards (Turner et al., 1995). Sampling was concentrated on Bowscar Quarry, Red Rock Quarry, a disused quarry at Salter Hill, and two natural outcrops at Force Mill and around Lacy's Cave.

All sampled settings have been subject to temperatures above 80 °C and the mineral transformations of possible precursor phases are expected to have started.

The differences in maximum burial depth, maximum temperature, present grain coating phases, and authigenic composition are summarized in table 4.1. Point count compilations for each sample can be found in the appendix (Tab. A.1 to A.4).

Table 4.1: Compilation of reservoir-related criteria of the sample set and overview of the main authigenic phases and their relative abundance.

	NE-Hannover	Dutch offshore	Bebertal	North England
Present TVD (m)	4264-4570	2479-3117	0	0
Max. Burial (m)	4264-4570	2479-3117	unknown	~3400
Max. temperature (°C)	162-185	87-115	200-250	~120
Coating phases	illite, hematite, chlorite, smectite	illite, hematite	illite, hematite	illite, hematite
Authigenic cement phases				
Quartz	+	+	+	+
K-spar	+	+	+	-
Kaolinite	-	+	-	+
Illite	+	+	+	+
Chlorite	(+) o	(o)-	o	-
Calcite	+	o	+	o
Dolomite (incl. Fe-Dolomite)	o	+	o	o
Anhydrite	+	+	+	o
Baryte	o	+	+	o

+ present
o minor amounts
- absent

4.4 Materials and methods

The delineation of dependencies governed by the depositional environment was based on core and outcrop description of the lithologies. Sample analysis was mainly performed on thin sections dyed with blue epoxy resin and assisted by SEM analyses. Samples from well F-J were all stained with Alizarin Red-S to differentiate carbonate cements. The transmitted light microscope is a Leica DMLP, equipped with an Invenio 5DII camera. The SEM analyses were performed with a JEOL JSM7000F at the Central Facility for Electron Microscopy (GFE), RWTH Aachen University, for SEM and SEM-EDX analyses at an acceleration voltage of 15kV and a working distance of approximately 10 mm.

The evaluation of modal compositions was based on point-counting (300 steps) on thin section samples. The Dutch samples (well F-J) are documented in GAPS (1982, 1983, 1984a, 1984b, 1994) and Winters (1989). The evaluation of coat coverages was performed for all samples on 50 grains per thin section (e.g. Taylor et al., 2004). The coverage of grains is determined as the surface of a grain, which is in contact with the intergranular volume and coated by a mineral phase other than syntaxial overgrowths. Grain contacts are omitted. Relative amounts of grain coat coverage are important input parameters for the Touchstone™ reservoir quality modeling approach. The measurements of grain sizes were performed for at least 100 grains per sample on a grid adjusted to the maximum observed grain size. Trask sorting was calculated from these measurements.

To show the impact of universal versus sample specific measurements, data was integrated into a diagenetic forward model. The diagenetic forward modeling software Touchstone ™ is widely used in the Exploration and Production Industry. Touchstone™ simulates the development of porosity, permeability, compaction, cement authigenesis, and mineral growth kinetics in dependence of burial depth, pressure and temperature (Makowitz et al., 2006; Taylor et al., 2010). Additionally, specific models allow the use of fibrous illite kinetics (Lander and Bonnell, 2010) and alternative quartz overgrowth kinetics (Lander et al., 2008). Besides 1D burial models, the whole available database of petrophysical and petrographic data was used. The coat coverages in the input database for correlation were adjusted to the universal arithmetic means deduced in this study.

4.4.1 Sediment Types

The following scheme was used to coherently re-interpret the depositional environment of sediments in core and outcrop material to assign a specific depositional environment for each of the analyzed samples. Accordingly classified to previous work in the siliciclastic Rotliegend (Antrett, 2011; Vackiner, 2011), we distinguished the following sediment types: eolian dune, dry sandflat, damp sandflat, wet sandflat, low-energy fluvial, lacustrine deposits, and lake margin deposits (Tab. 4.2).

Eolian dunes represent 102 samples in this study, dry sandflat deposits make up 69 samples. Damp and wet sandflats pose for 29 and 5 samples respectively. Fluvial deposits were sampled 33 times and lake margin deposits were sampled 3 times.

Cores from NE-Hannover include all studied depositional environments, but low-energy fluvial deposits were not sampled. Cores from the Dutch offshore cover the full range of depositional environments (Fig. 4.2). Sampling in Bebertal and North UK was concentrated on eolian dune and low-energetic fluvial deposits.

Table 4.2: Overview of criteria to distinguish the depositional environment from core and outcrops.

Depositional Environment	*Characteristic*	*Reference*
Eolian Dune	well-bedded, fine to coarse grained sandstone, unimodal dip, bimodally sorted, grain fall- and grain flow laminae	Anderson (1988), Fryberger et al. (2011)
Dry Sandflat	well bedded or homogeneous, fine to coarse grained sandstone, bedding generally less than 5°	Mountney and Jagger (2004)
Damp Sandflat	fine to medium grained sandstone, silt and clay content up to 20 %, wavy lamination, can contain soft sediment deformation	Amthor and Okkerman (1998), van den Belt and van Hulten (2011)
Wet Sandflat	very-fine to fine sandstone, silt and clay content between 20-50 %, poor sorting, wavy lamination, can contain soft sediment deformation	Amthor and Okkerman (1998), van den Belt and van Hulten (2011)
Low-energetic Fluvial	clay to coarse sandstone or conglomerate, can contain rip-up clasts in rather homogeneous sandstone matrix	Fryberger et al. (2011)
Lacustrine	100% structureless or laminated shale	Fryberger et al. (2011)
Lake Margin	irregular cm-scale intercalations of poorly sorted, structureless sandstone in shale, grading into lacustrine deposits	Antrett (2011)

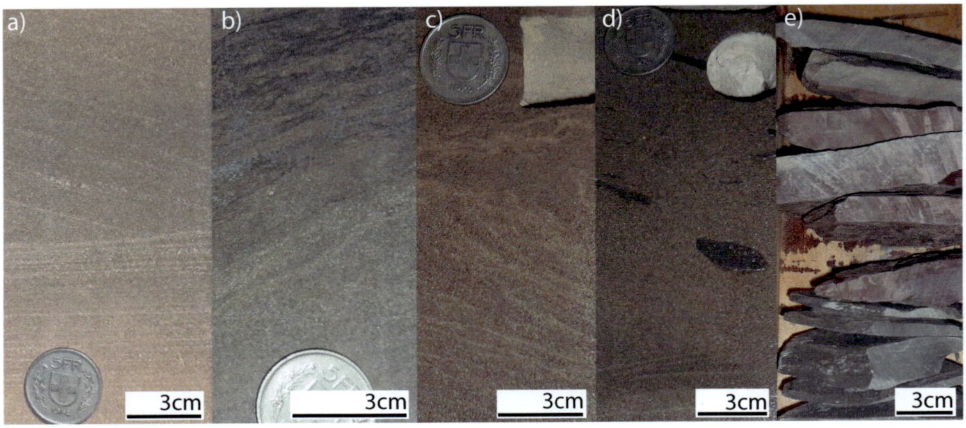

Figure 4.2: Lithological associations examples from the Dutch offshore region. a) Top: pin stripe laminated, cross bedded dune sandstone, bottom: horizontally laminated dry sandflat containing deflation layers. b) Wavy lamination in clay-richer damp sandflat environments. c) Irregular lamination in clay rich (>50%) wet sandflat deposits. d) fluvial deposits occasionally containing clay rip-up clasts. e) Lacustrine deposits consisting of occasionally finely laminated clay. Where the lithologies are bleached the clay layers show reaction horizons towards the over- and underlying sandstones.

4.5 Results

4.5.1 Depositional environment

The grain coating coverage of all samples ranges from very low coverages (~5 %) to very high coverages (~97 %) (Fig. 4.3). The grain coat coverage for the dune environment varies from 13.6 % to 97.2 %. The range of grain coat coverages for dry sandflat environments is between 5.4 % to 97.2 %. Damp sandflat deposits cover 22.9 % to 91.2 %, whereas wet sandflat deposits extend from 28.9 % to 69.4 %. Fluvial deposits exhibit coverages between 19.1 % to 83.3 % and lake margin deposits cover 30.8 % to 65.1 %. A clear delineation is not possible with medians (lines in Box-Whisker-Plot, Fig. 4.3 a, Tab. 4.3) and arithmetic means (small boxes is Box-Whisker-Plot, Fig. 4.3 a, Tab. 4.2) which show little variation and not much separation. Additionally, damp and wet sandflat, as well as lake margin deposits are characterized by larger amounts of detrital clay in the pore space and as layers, as opposed to samples from the other depositional environments. This impacts the evaluation of clay coat coverages and reduces available pore space for quartz cementation. Additionally, no clear correlation with the sampling area could be established, since they all cover the observed range (Fig. 4.3 b).

Since no consistent accumulation of coatings could be established for a certain depositional environment, the potential impact is omitted in the following examinations.

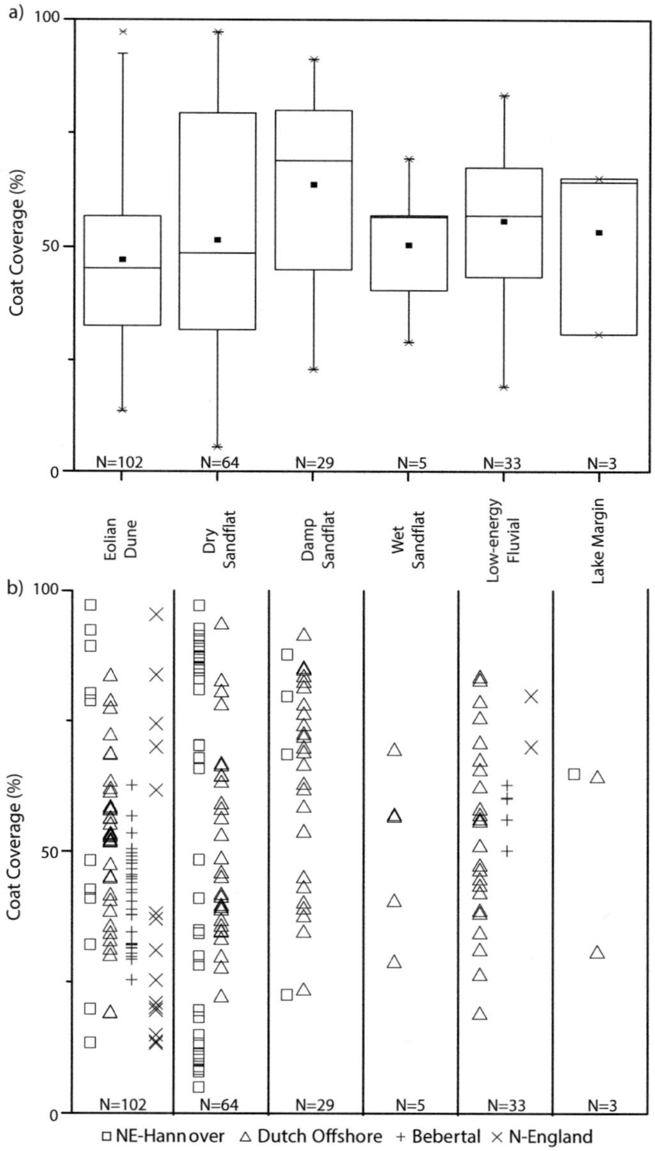

Figure 4.3: a) Box-Whisker plots of the coat coverages plotted for the studied depositional environments. The compilation highlights the spread of coat coverages. The median of the samples is given by the lines inside the box, the arithmetic mean is given by the solid black rectangles. Dashes at the top and bottom give the 0% and 100% data, x marks the 1% and 99% data points. b) The raw data split up by the core areas. The overlap in data does not delineate a consistent trend assignable to any core area.

Table 4.3: Compilation of the arithmetic mean, median, minimum and maximum values for the evaluation of coat coverages of all studied samples.

	Eolian Dune	*Dry Sandflat*	*Damp Sandflat*	*Wet Sandflat*	*Low-energy Fluvial*	*Lake Margin*
Arithmetic Mean	47.3	51.7	63.6	50.4	55.3	53.4
Standard Deviation	19.4	27.5	20.0	15.8	16.7	19.6
Median	45.3	48.6	68.9	56.6	56.2	64.2
Minimum	13.6	5.4	22.9	28.9	19.1	30.8
Maximum	97.3	97.2	91.3	69.4	83.4	65.1

Additionally, the influence of the position within an eolian dune unit was studied on subsets of outcrop samples from Schwentesius Quarry, Germany and Bowscar Quarry, N-England (Fig. 4.4). Samples from above a bounding surface, which correspond with the base of an eolian dune, and samples from below a bounding surface, corresponding to the top of the lower eolian dune unit, have been evaluated to derive a possible accumulation. The set is completed by samples from the bounding surface itself.

A slight decrease in coat coverage below the bounding surface as opposed to above the bounding surface could be delineated from the Bebertal sample set (+ in Fig. 4.4). The coat coverages on average (arithmetic mean) do not differ enough from below to atop the surface, to exert a significant change in cementation. The coat coverage from the N-England sample show no difference neither below, nor above the bounding surface (x in Fig 4.4). No generally observable accumulations towards higher or lower clay coat coverages at the eolian dune base or the eolian dune top can be seen.

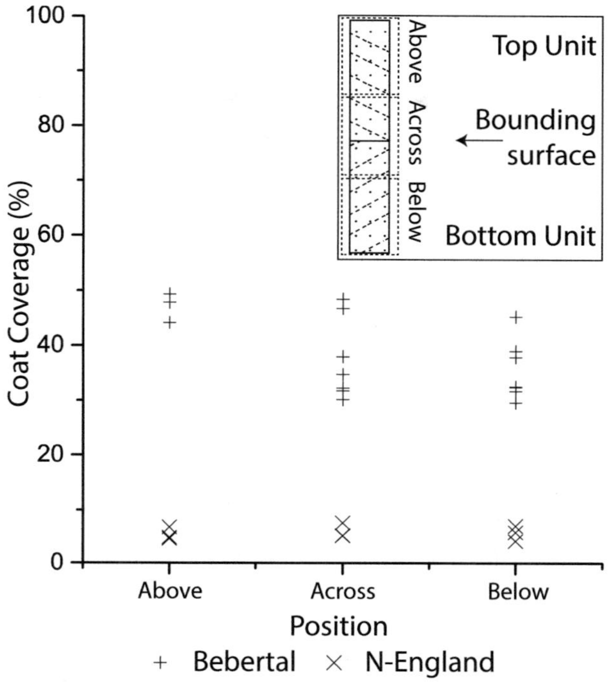

Figure 4.4: Coat coverages of the outcrop sample sets from Bebertal and N-England. No significant differences can be observed at the eolian dune base as opposed to the top of the underlying eolian dune. Inset shows sample position relative to bounding surface marking the change between two eolian dune bodies.

4.5.2 Grain size and sorting

A correlation of better coated finer grains seems possible if the dataset is plotted in Box-Whisker plots in dependence of grain size classes (Fig. 4.5). This visualization suggests that finer grains are, on average (arithmetic mean, black boxes in Fig. 4.5), better coated than coarser grains. However, the spread of observed coat coverages for the medium and fine grained portions is nearly the same.

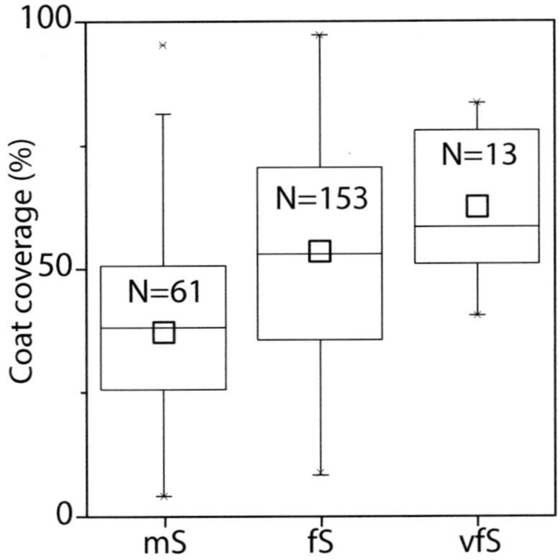

Figure 4.5: Box-Whisker plot of the distribution of observed grain coat coverages in the different grain size classes of medium (mS), fine (fS), and very fine grained sandstones (vfS). Although the arithmetic means and medians indicate a preference of higher coat coverages in finer grained sandstones, the spread of coverages in medium and fine grained sandstones is the same.

On the scale of each sampling area, grain coat coverages weakly correlate with average grain size (arithmetic mean), with smaller grains being more well covered than larger grains (Fig. 4.6). Samples from NE-Hannover reservoirs show poorest decrease, i.e. the amount by which the coat coverage is reduced depending on the grain size (Fig. 4.6 a, coefficient of correlation $R^2 = 0.04$).

The sample set from the Dutch offshore shows a large scatter of values (Fig. 4.6 b). The quality of the slight correlation of finer grain size versus better grain coat coverage is impacted by this scatter ($R^2 = 0.02$). A weak correlation ($R^2 = 0.1$) was attained for the Bebertal sample set (Fig. 4.6 c) showing higher coat coverages with decreasing average grain size. However, the observed coat coverages only cover a rather narrow range from 25 to 67 %. Here, the fluvially deposited samples, on average, show larger grain coat coverage than the eolian dune samples.

In the sample set from N-England (Fig. 4.6 d) the grain size exerts a weak negative correlation with grain coat coverage ($R^2 = 0.24$). Additionally, the fine sand portion is generally better coated than the medium and coarse sand fraction, especially in very characteristically pinstripe laminated eolian dune sandstones (Fig. 4.7). The medium to coarse sand fraction only contains pigmented hematite rims (Fig. 4.7 a) whereas illite is accumulated in the fine sand parts of the sample (Fig. 4.7 b). Their grain rimming character is best visualized by transmitted light microscopy. This local factor was not observed samples from other areas in this study.

The additional color coding of depositional environment bears no consistent basis for additional correlations (Fig. 4.6).

Regardless of sorting, coat coverages consistently cover a range from poor to well coated and do not concentrate (Fig. 4.8). A separation of clusters is not evident. Only the range of coating coverage seems to decrease with a decrease in sorting.

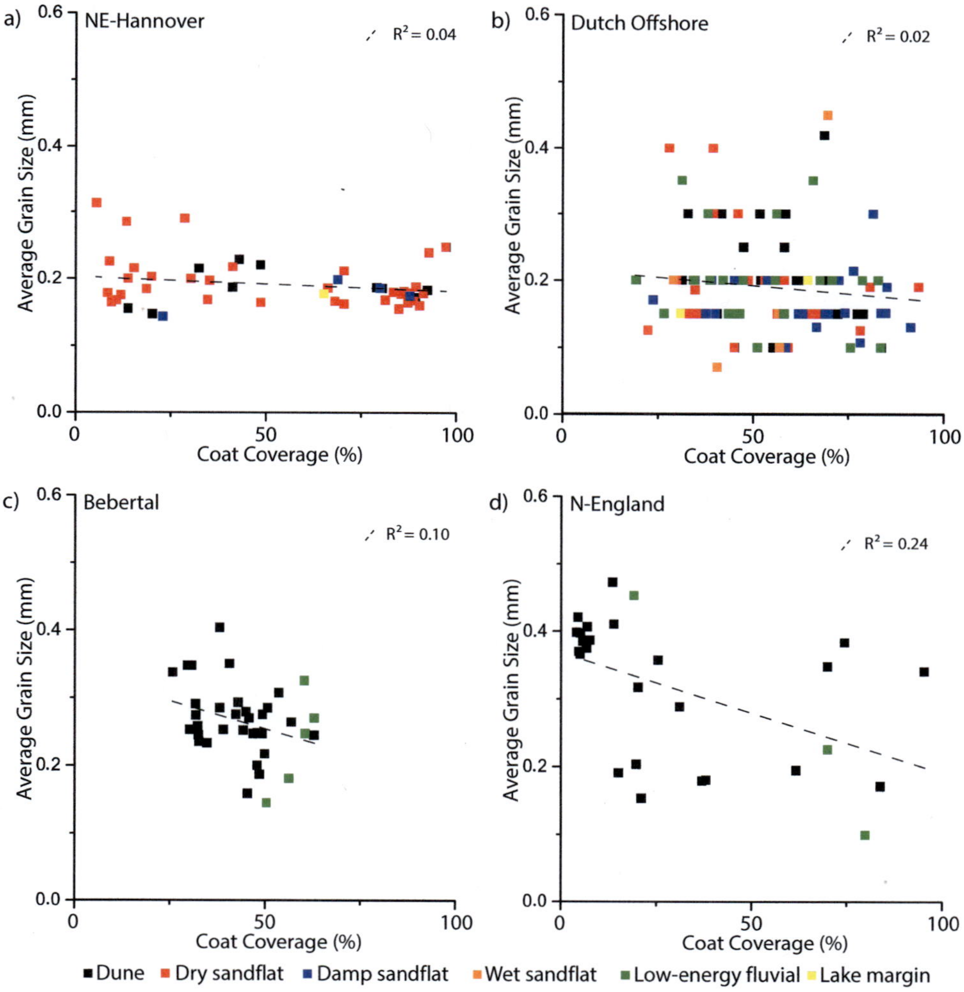

Figure 4.6: a-d) Correlation of coat coverage versus the average grain size in mm. No clear and consistent trend can be seen, as the data scatters across the range of grain sizes and coat coverages. The color code represents the depo-sitional environments.

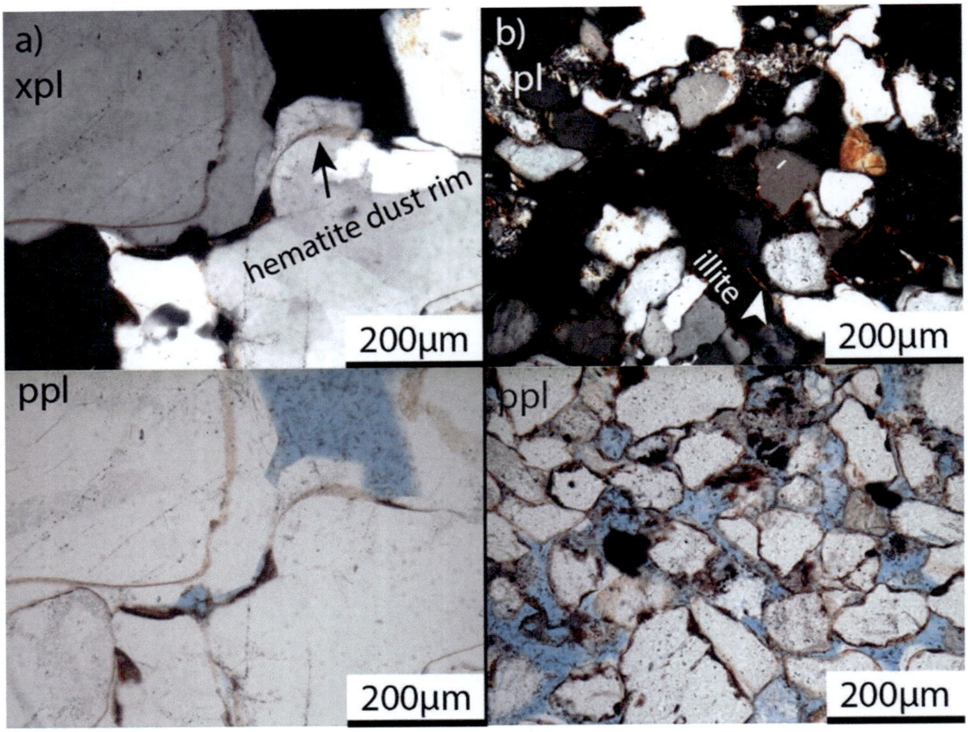

Figure 4.7: a) Pigmented hematite dustrim with occasional illite accumulations in indentations from sample BQ_2. The quartz grains are completely overgrown. b) Finer grained laminae of the same sample showing more complete coatings of illite (arrow in xpl image). Syntaxial quartz overgrowth cementation in this part of the sample is inhibited. (magnification of Fig. 3.5)

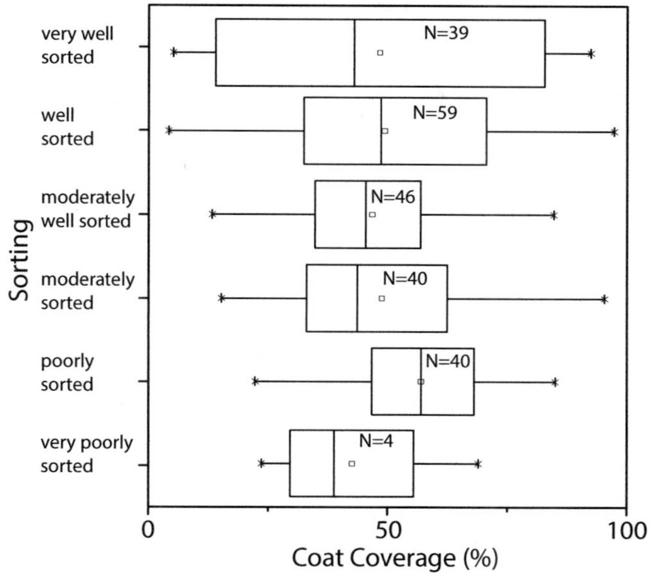

Figure 4.8: Correlation of Trask sorting and coat coverages. The baseline of well to very well sorted sandstones already covers the whole range of determined grain coat coverages. Even for the less well sorted sandstones, no clear or consistent trend is visible.

4.5.3 Detrital grains

Significant differences in detrital composition are related to the location (Fig. 4.9). Samples from NE-Hannover are mainly classified as subarkoses, Dutch samples as arenites and sublitharenites, Bebertal samples as sublitharenites, feldspathic litharenites, and litharenites, and N-England samples as quartz arenites to subarkoses (classification after Folk (1980), Fig. 4.9). A clear relationship of detrital composition and the depositional environment cannot be identified (signatures in Fig. 4.9).

However, the presence of rock fragments and especially of volcanic rock fragments can influence the development of clay coatings, i.e. chlorite rims (Anjos et al., 2000; Anjos et al., 2003; Bahlis and De Ros, 2013).

The only sample set comprising chlorite coatings was from NE-Hannover. Depending on the well, those samples contain chlorite and smectite-chlorite coatings as well as illitic and hematitic coatings and a wide variety of coat coverages. The occurrence of smectite-chlorite and chlorite coatings is restricted to the eastern wells A, C, and E. Well E also contains hematite rims. All other samples contained either illite coatings or pigmented hematite dust rims, but no chlorite coatings. The chlorite was identified by EDX spectra showing silicon, magnesium, iron, aluminium, and oxygen and by weak greyish interference colors in cross polarized lights (xpl) and a weak green color in plane polarized light (ppl). The chlorite either appears tangential on grain surfaces (Fig 4.10 a), where smectite was not detected, or as small rosettes (Fig. 4.10 b), where it grew on tangential smectite coatings, also partly replacing the smectite.

The evaluation of the correlation of unstable volcanic rock fragments including their alterations and chloritic grain coatings yielded no consistent result. All samples contained volcanic rock fragments (Fig. 4.11), here mostly basaltic fragments, characterized by plagioclase laths, occasionally in an opaque matrix. Only samples of 3 of 5 wells (A, C, E) contained chloritic coatings (Fig. 4.12 a). The in-situ alteration of volcanic rock fragments towards chlorite however, was also observed in samples from wells B and D, containing hematite and illite coatings (Fig 4.12 b). Consequently, no consistent impact of volcanic rock fragments could be observed.

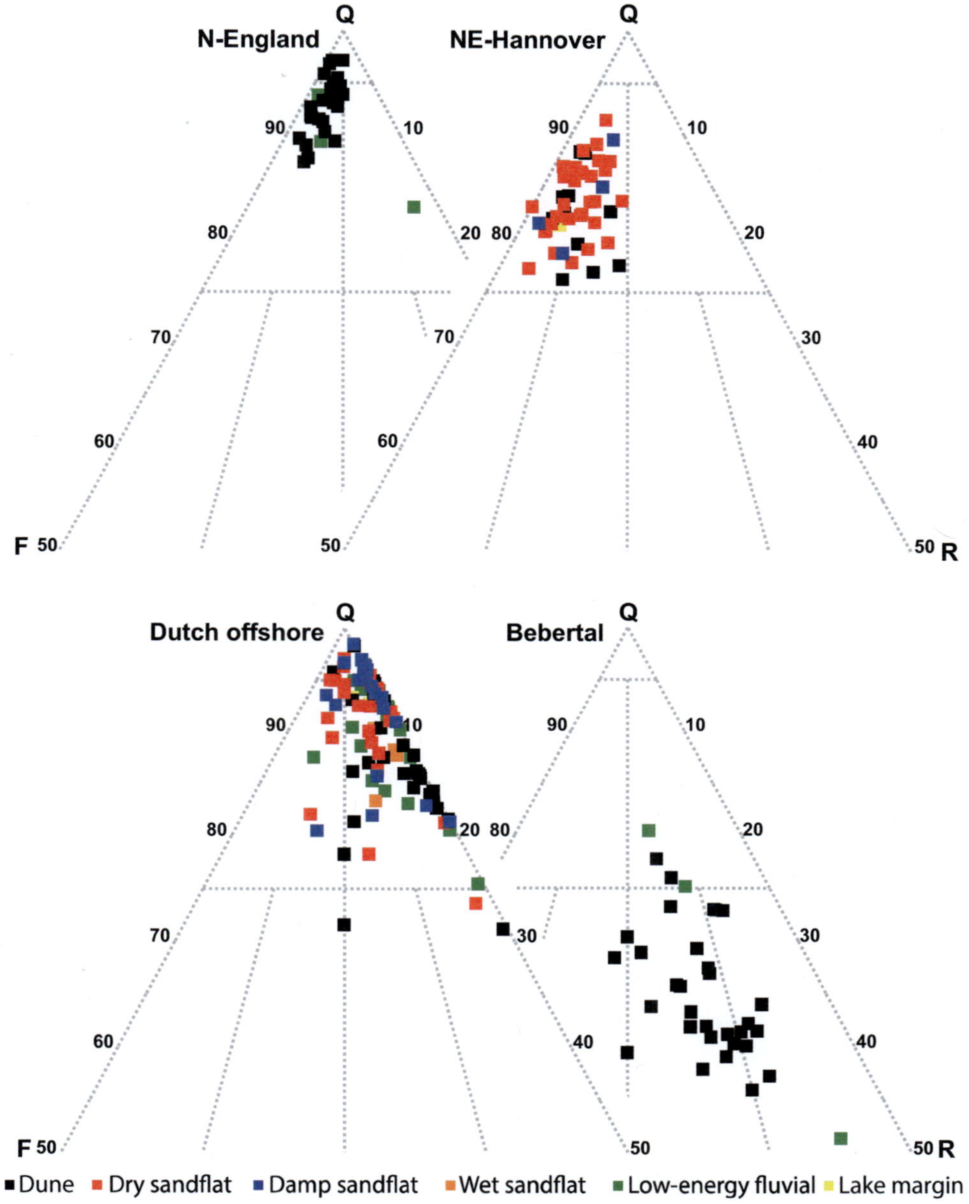

Figure 4.9: QFR plots for the different study areas showing the detrital composition in dependence of their depositional environments. No clear accumulations are visible across the Permian strata studied in this study. The QFR plots only show compositions above 50 % quartz.

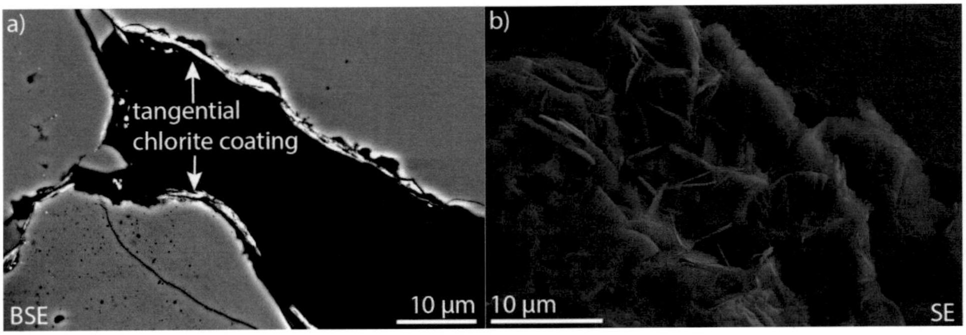

Figure 4.10: Scanning electron microscope images of different chlorite coating morphologies in samples from Well E. a) Tangential chlorite coating in back scattered electron image (BSE), possibly replacing an early smectite precursor. b) Chlorite rosettes growing on top of a smectite precursor in secondary electron image (SE).

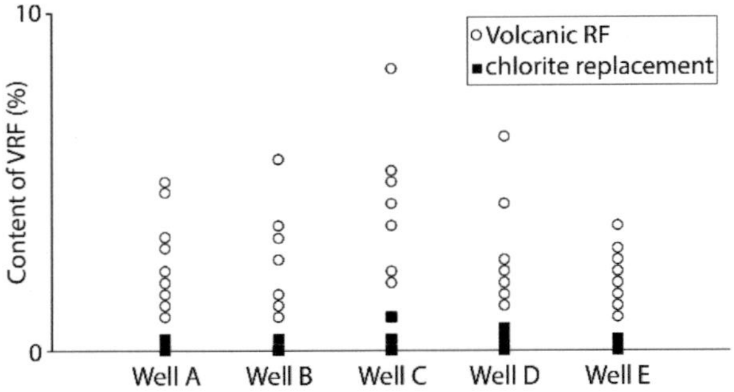

Figure 4.11: Compilation of point-counted results of the volcanic rock fragments. All samples contain volcanic rock fragments in varying amounts. Samples from all wells contain the replacement of volcanic rock fragments by chlorite.

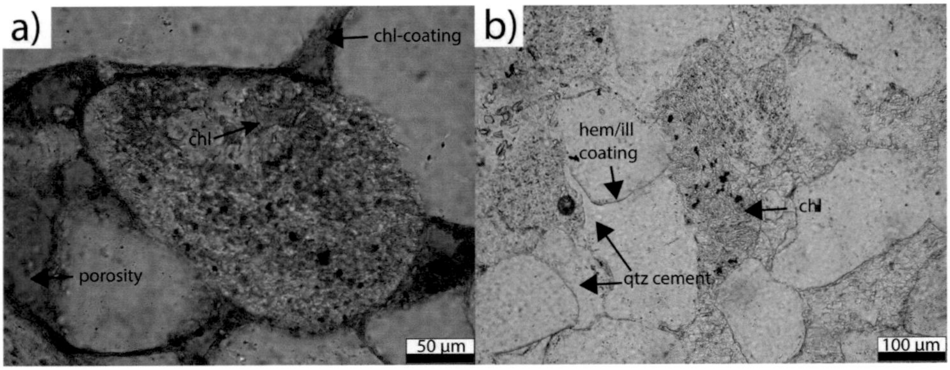

Figure 4.12: a) Chlorite replaces volcanic rock fragments in sample from well E. The sample is mainly characterized by chlorite coatings. b) Chlorite replaces plagioclase-rich volcanic rock fragment in sample from well D. The coatings in this sample consist of hematite pigments and minor amounts of illite.

4.6 Discussion

4.6.1 Factors influencing the formation of grain coatings in fluvio-eolian systems

The factors controlling the occurrence and extent of grain coat coverages have been controversially discussed in several published studies. This chapter compiles and determines the main controlling factors as derived from published literature. Some of the models directly contradict each other, i.e. sometimes a favoring effect is ascribed to finer grains, other times it is ascribed to favor a more complete grain coat coverage in coarser grains (Walker, 1979; Bloch et al., 2002; Ajdukiewicz et al., 2010; Darbha et al., 2012). Additionally, the timing of illite or chlorite formation during early diagenesis or burial diagenesis is essential (e.g. Liewig and Clauer, 2000; Anjos et al., 2003). Moreover, the interaction of numerous processes prevents the straight forward prediction (Tab. 4.4).

Adding the influence of different formation fluids during diagenesis and variable maximum burial depths, the variability in diagenetic overprints will have an impact on the preservation of coatings from deposition to present day. Even on a small km scale, as the presented data from NE-Hannover, diagenesis and depositional controls act in concert and require the attention of a petrographer. These effects may not be neglected during the analysis and prediction of reservoir quality.

Additionally, the formation of pore lining clay phases during burial diagenesis can be a contributor to the inhibition of syntaxial quartz cementation in these settings. Fibrous illite, which appears pore lining, forming a radial rim on quartz grains, can inhibit later burial diagenetic quartz cementation. These radial illite phases can nucleate on smectitic/illitic substrates (Gaupp et al., 1993), but are also documented to form without a substrate (Liewig and Clauer, 2000). This later diagenetic phase contributes to the preservation of porosity, but is not directly governed by the depositional environment.

Table 4.4: Compilation of controlling factors on the presence and extent of grain coatings.

Name	*Description*	*Reference*
Grain size	Coarser grains have less specific surface area, require less clay material to be well coated than finer grains.	Bloch et al. (2002)
Grain size II/ Dune mobility	Smaller grains are able to better retain coatings during transport and abrasion in mobile eolian dune settings, due to a more rugose surface	Ajdukiewicz et al. (2010), Walker (1979), Darbha et al. (2012)
Meteoric waters	Periodic phases of wetting and drying to form most complete coats	Ajdukiewicz et al. (2010)
Ground water	High salinity of groundwater during early diagenesis may result in authigenesis of chlorite coatings	Ajdukiewicz et al. (2010)
Wind direction	Impact on grain size or texture of sediment and following alteration of coat properties plus source for airborne dust and transport medium for dust before settling	Ajdukiewicz et al. (2014), Esch et al. (2008)
Source area for inherited rims	Inherited rims from ephemeral fluvial deposits	Esch et al. (2008)
Rock fragments	Burial diagenetic alteration of ferromagnesian grains to chlorite can form coatings	De Ros et al. (1994), Remy (1994), Anjos et al. (2003), Bahlis and De Ros (2013)
Authigenesis	Radial, pore lining illite in burial diagenesis, as result of hyperthermal event.	Liewig and Clauer (2000)
Authigenesis II	Radial illite found in early near-surface diagenesis in phreatic zones	Ajdukiewicz et al. (2010)
Early cementation	Early pore filling carbonate and sulfate cementation in arid setting might prevent the infiltration, illuviation, and formation of early and late clay coats	Morad et al. (2000)
Mesogenesis	Final coating characteristics develop during recrystallization of precursor mineral phase	Aagaard et al. (2000)

4.6.2 Depositional environment

The studied depositional environments of Rotliegend sandstones in the Southern Permian Basin do not correlate with the coverage of detrital grains by coatings. However, multiple dependencies of reservoir properties on the depositional environments are evident from literature (Ramm, 2000; Elias et al., 2004; Khalifa and Morad, 2015; Marchand et al., 2015). Although the arithmetic means of the data show a slight separation between depositional environments (solid black boxes in Fig. 4.3 a), they will result in incorrect associations since the extremes are responsible for reservoir quality anomalies.

No consistent dependencies can be established for these Rotliegend samples from the Southern Permian Basin, most likely due to differences in the source areas, in climate, water availability and the following diagenetic pathway. It is still a matter of discussion if an incomplete coat coverage during early diagenesis might act as nucleation for further mineral growth as proposed by Seeman (1982), or if a complete coverage would be necessary upon deposition before significant diagenetic changes occur. Smallest changes in the climate do seem to exert a major control on the authigenic development of these sandstones, regardless of depositional environment. Variations of detrital compositions might additionally impact the later diagenetic alteration (see below). However, a correlation of grain coat coverages to certain depositional environments as opposed to others can clearly be discarded for the studied fluvio-eolian Rotliegend samples. This highlights the need for area and sample specific analysis for reservoir quality modeling and prediction. Variations may even become apparent within one location as shown by the samples from NE-Hannover. Additionally, samples from outcrops in Bebertal and N-England which show an increase in grain coat coverage in sandstones, were closely associated with fluvially deposited sands. These local dependencies cannot be transformed into a universal dependency.

The position of samples in respect to the base or topmost part of sedimentary bodies around bounding surfaces does not affect a change in coating characteristics in the sample sets Bebertal and N-England. Variations in duration of infiltration of fines in early diagenetic fluids and their concentration and composition was previously reported as a possible reason for observed differences around bounding surfaces (Fischer et al., 2012). Our database shows no clear correlation of this process being effective.

4.6.3 Grain size and sorting

The grain size has a small impact on the grain coat coverages. Due to the fact that in coarse-grained sandstones the grain surface area is smaller than in fine-grained sandstones, less material is required in coarse-grained sandstones to achieve a more complete grain coat coverage (Bloch et al., 2002). However, in the studied dataset, the coat coverage may increase with decreasing grain size, although correlation coefficients (R^2) are poor (Fig. 4.6). The exception is the strict bimodal lamination in sandstones from the eolian dune environment in the generally less coated samples from the N-England set (Fig. 4.7). Here, the fine sand portion was generally more completely coated as opposed to the intercalated medium to coarse sand fraction. Differences could be due to capillary retention in finer grained media with a smaller pore size (Smith et al., 1930), influencing the illuviation and distribution of clay minerals. Illuviated clay minerals could thus accumulate in these regions during evaporation. Additionally, a process governed by the grain size might be the abrasion of emplaced precursory clay phases. Finer grains are discussed to show less abraded grain coats as opposed to coarser grains (Ajdukiewicz et al., 2010). Abrasion would additionally leave remnants of clay coats in indentations in larger grains (cp. Fig. 4.7). This process is held responsible for observed differences of coat coverage in laminated wind-ripple sands. However, because of the complete absence of grain coating clays and no clear indication for eolian dune mobility (other than said criteria), we encourage to add the possibility of capillary retention of water shortly after deposition to any genetic model of grain coat formation.

The sorting was evaluated from the grain size measurements. Universal trends cannot be delineated since most of the sorting classes cover a broad range of observed grain coat coverages and no separation is visible.

4.6.4 Detrital grains

Variability in detrital composition is strongly related to the location of the studied samples. Samples from NE-Hannover are rich in feldspar and contain some rock fragments, while Dutch samples are poor in feldspar, mainly related to dissolution and alteration to kaolinite. Bebertal samples are very rich in rock fragments and still contain considerable amounts of feldspars. Samples from N-England are generally poor in feldspar and rock fragments, mainly due to dissolution and alteration of feldspars and more generally the different source area, recycling the Millstone Grit Series (Waugh, 1970b). For the settings near Hannover, Bebertal and the Netherlands, the source areas are different areas of the Variscan orogenic belt (e.g. Verweij, 2003).

In the Rotliegend dataset, no direct link between the alteration of volcanic rock fragments and grain coat coverages can be identified. Adjacent wells with similar intensities of volcanic rock fragment alteration exhibit strong variations in grain coat coverages and presence of grain coating chlorite. Alterations of volcanic rock fragments towards chlorite are observed in samples from all wells in the NE-Hannover area (see Fig. 4.11, 4.12). The presence of chloritic coatings in samples from wells A, C, and E and their absence in samples from well B and D, only containing minor illite and hematite, can be explained by the incomplete coat coverage of precursory grain coatings. The incomplete coatings are readily overgrown by quartz cement thus isolating the precursory and incomplete coatings and grain surfaces in wells B and D from further alteration and emplacement. The overgrown coatings are isolated from the pore volume and no available cations are further incorporated during mineral transformation.

The alteration of smectite to chlorite is also reported to favor the authigenic replacement by chlorite as a grain rim in presence of altered volcanic rock fragments (Bahlis and De Ros, 2013). Where the precursory coat coverage is larger, pervasive quartz cementation is inhibited. Resulting direct exposure of precursory coating phases to the connected pore volume allow the alteration to chlorite in the presence of cations released into solution by the alteration of volcanic rock fragments. This process could explain the observed differences in coat chemistry and coat coverage from wells A-E. The presence of smectite as a substrate for chlorite growth could be shown in this sample series. This additionally indicates the independence of grain coat formation from the depositional environments in this setting.

The direct alteration of volcanic rock fragments towards chlorite grain coatings during early diagenesis (Bahlis and De Ros, 2013) can be disregarded for samples from NE-Hannover. However, their role as cation donors during chlorite authigenesis and smectite replacement should be considered.

4.6.5 Diagenetic modeling

We tested the reliability of universal grain coat coverages in comparison to measured sample-specific analog data from cores. Here, we present the impact of the use of universal coat coverages based on the depositional environment. In an existing diagenetic model, the measured grain coat coverage of each sample has been exchanged for the arithmetic mean of the respective depositional environment (Fig. 4.3 and Tab. 4.2). They have been incorporated into the diagenetic model of the five wells from NE-Hannover. The calibrated model based on measured coat coverages reproduces the measured amounts of syntaxial quartz cement well and rarely exceeds 4 % deviation (Fig. 4.13 a). However, the universal coat coverage as established for the depositional environment results in an overprediction of quartz cement values as compared to the measured amounts (Fig. 4.13 b). This is especially visible in wells A, C, and E, where measured coat coverages are much larger than the universal values. Additionally, the scatter of data is broader around the 1:1 line, which cannot be reversed by adjusting other modeling parameters without changing the input data.

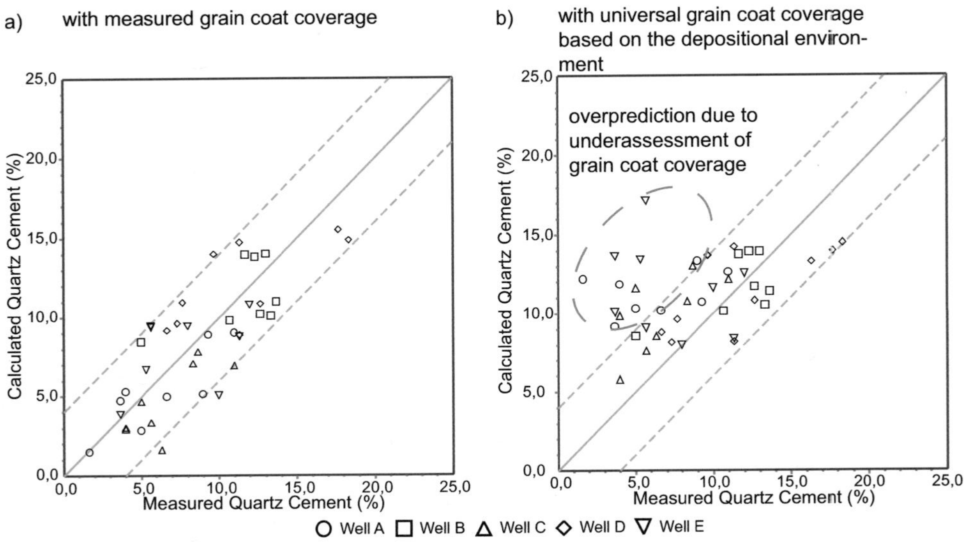

Figure 4.13: a) Calibration plot of measured quartz cement versus calculated quartz cement. The values scatter around the 1:1 diagonal and do not deviate more than 5% from the measured values. b) Calibration plot based on the universal coat coverages established during this study. The values deviate from the measured values especially for samples from wells A, C, and E. These samples show the largest coat coverages in the study.

The mismatch in diagenetic modeling results versus measured authigenic quartz cement abundance highlights the need for accurate petrographic analysis and evaluations of grain coatings to reproduce the measured amounts of authigenic quartz cement. The deviation is especially striking in samples from wells A, C, and E (Fig. 4.13 b), which contain the largest coat coverages. In an E&P industry approach, this mismatch would ultimately impact the prediction of reservoir quality in undrilled fields. The amount of quartz cement will be overpredicted, thus porosity will be underpredicted while the simulation still aims to match intergranular volume (IGV, cement + intergranular porosity) values. It thus results in the evaluation of poorer reservoir qualities than those that might be found. To prevent these deviations, it is imperative that the reservoir quality is estimated based on nearby analog wells in this Rotliegend setting.

4.7 Conclusions

Grain coat coverage cannot be directly linked to the depositional environment, sorting, or presence of detrital volcanic grains for Rotliegend sandstones across the Southern Permian Basin (NE-Hannover, Dutch offshore, Bebertal). Even on a small (km-) scale (e.g. NE-Hannover) no universally applicable controls can be observed. Locally, the best correlation was achieved between grain size and coat coverage. But the overall poor correlation and the large data scatter do not allow the use of these parameters as a universal predictive tool for the Rotliegend. Hence, no universal dependencies can be derived for grain coating prediction for diagenetic modeling in Rotliegend sediments.

The impact of grain size varies from one location to another, due to the varying superposition effect by other controlling processes and later diagenetic overprint (e.g. dissolution, nucleation). This includes winds, erosion, clay availability, pore water chemistry, and meteoric water. This investigation strongly confirms that the emplacement is not only controlled by a single process, but by a complex interaction of numerous parameters.

Although the arithmetic means of the correlations with the depositional environment, grain sizes, and sorting suggest a correlation with the grain coat coverage, the variations and extremes of grain coat coverage are responsible for reservoir quality heterogeneities. Thus, they must be considered in reservoir quality assessment and modeling. The mismatches of reservoir quality modeling on reservoir evaluation are too severe to determine universally applicable dependencies of grain coat coverages from this Rotliegend database.

5 Evolution of small-scale flow-barriers in German Rotliegend siliciclastics

5.1 Abstract

Many siliciclastic reservoirs contain mm-scale diagenetic and structural phenomena affecting fluid flow. We identified three major types of small-scale flow-barriers in a clastic Rotliegend hydrocarbon reservoir: cataclastic deformation bands, dissolution seams and bedding-parallel cementation. Deformation bands of various orientations were analyzed on resistivity image logs and in core material. They are mainly conjugates and can be used to validate seismically observable faults and infer subseismic faults. Bedding-parallel dissolution seams are related to compaction and postdate at least one set of deformation bands. Bedding-parallel cementation is accumulated in coarser grained layers and depends on the amount of clay coatings and fine grained layers adjacent to dissolution seams.

Permeability data are related to petrographic image interpretation to visualize the impact of flow barriers on reservoir heterogeneity. Transmissibility multiplier calculations indicate the small efficiency of the studied deformation bands on flow properties in the reservoir. Deformation bands reduce the host rock permeability by a maximum of two orders of magnitude. However, host rock anisotropies are inferred to reduce the permeability by a maximum of four orders of magnitude. The relative timing of these flow-barriers as well as the assessment of reservoir heterogeneities are the basis for state of the art reservoir prediction modeling.

5.2 Introduction

Deformation bands are zones of localized deformation in granular media and are frequently reported from siliciclastic rocks and limestones (e.g. Antonellini and Aydin, 1995; Fossen et al., 2007; Legler and Marchel, 2008; Wennberg et al., 2013). Generally three end members of deformation bands are kinematically classified as shear bands, compaction bands and dilation bands (Fossen et al., 2007). The orientation of the localization plane with respect to the principal stress orientation and the deformation band type differs for these three end members (Bésuelle and Rudnicki, 2004, their Fig. 5.16). Shear localization forms at low effective stresses, compaction at higher effective stresses and dilation is linked to decreasing effective pressures (Wong et al., 1997; Bésuelle and Rudnicki, 2004). Low porosity and small grain size exert an influence on the physical process of strain localization and will result in an increased magnitude of the compactive yield strength (Wong et al., 1997; David et al., 2001; Schultz et al., 2010).

The wide range of varying microstructures in deformation bands is reflected by the diverse terminology, addressing the mineralogical composition (e.g. the incorporation of clay minerals into the deformation band by shearing results in phyllosilicate bands), the kinematic or physical processes of formation (e.g. dilational deformation bands, cataclastic deformation bands and compaction bands) or hybrids of different mechanisms (e.g. shear enhanced compaction bands). Cataclastic deformation bands are frequently observed around larger normal faults in soft sediment and weakly lithified rocks and display the incipient stage of faulting and strain hardening (Antonellini and Aydin, 1995; Fossen, 2010; Ballas et al., 2012; Soliva et al., 2013). They are reported to have no or only a very little impact on reservoir properties due to their spatially limited geometry and the limited interconnectivity in three dimensions (e.g. Fossen and Bale, 2007), while their clustering may act as a permeability barrier which reduces fluid flow up to four orders of magnitude (e.g. Saillet and Wibberley, 2013).

The geometry, evolution and different scaling relationships of deformation bands has been studied in many outcrops (e.g. Antonellini and Aydin, 1994; Fisher and Knipe, 2001; Ogilvie and Glover, 2001; Schultz and Fossen, 2002; Davatzes and Aydin, 2003; Eichhubl et al., 2004; Olsson et al., 2004; Parry et al., 2004; Eichhubl and Flodin, 2005; Sternlof, 2006; Ahmadov et al., 2007; Rotevatn et al., 2007; Johansen and Fossen, 2008; Schultz et al., 2008; Guo et al., 2009; Kolyukhin et al., 2009; Fossen, 2010; Fossen et al., 2011; Ballas et al., 2012; Chemenda et al., 2012; Exner et al., 2013; Nicol et al., 2013; Saillet and Wibberley, 2013; Schueller et al., 2013; Schultz et al., 2013; Soliva et al., 2013; Awdal et al., 2014; Torabi, 2014). While deformation bands often strike parallel to extension faults, local stress perturbations around faults, faults splays and in relay ramps may result in multiple sets of deformation bands (e.g. Antonellini and Aydin, 1995; Johansen et al., 2005; Rotevatn et al., 2007).

In this study we analyze the apparent permeability anisotropy of moderately to very well sorted Rotliegend sandstone cores and the influence of cataclastic deformation bands and bedding anisotropies on fluid flow. From microstructural analyses we derive a scaling relationship considering the width of cataclastic deformation bands and the grain size of the undisturbed host rock, which is correlated with the apparent permeability data. Data mining tools are used to better display variations of deformation bands in a reservoir. Finally, transmissibility calculations in relation to deformation band frequency and the assessment of deformation bands from core material are critically discussed.

5.3 Geological setting

The Rotliegend gas-producing reservoir is situated in a part of the N-S oriented fan shaped Graben setting in the North German Basin (NGB) in the area between Bremen, Hamburg and the Elbe river (Kayser, 2006). It is part of the Southern Permian basin and accumulated thick successions of Upper Rotliegend continental siliciclastics and evaporites. The Permian horst-and-graben structure (Fig. 5.1) was affected by Triassic to Cretaceous extension and formation of Zechstein salt structures (Vackiner, 2011). The geological evolution was followed by Upper Cretaceous N-S oriented compression, concurrent inversion and further anticlockwise stress rotation to the recent stress field in Late Eocene to Miocene (Lohr et al., 2007; Kley et al., 2008; Legler and Marchel, 2008). The sediments have mainly been deposited in an arid to semiarid desert environment and the red to grey sandstones fall into the field of lithic subarkoses to subarkoses (Legler and Marchel, 2008).

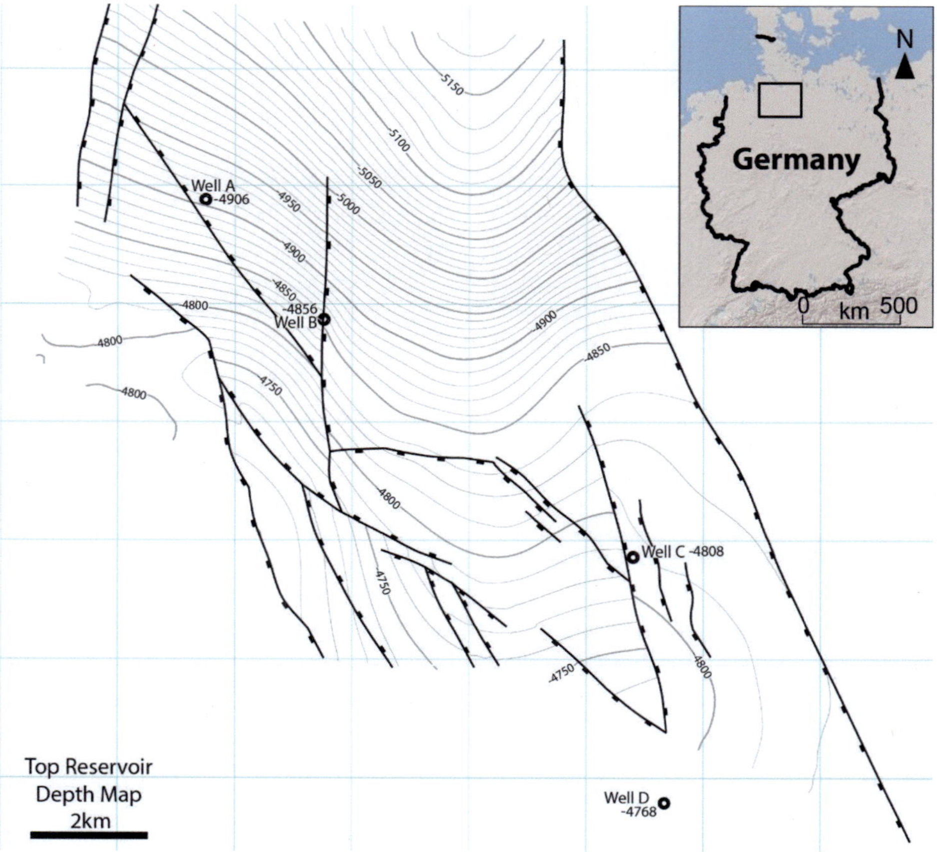

Figure 5.1: Location of the study area in northern Germany (right) and structure and depth map of the top of the reservoir unit including well locations (left).

The main reservoir target for exploration and production is the Upper Rotliegend Havel Subgroup (266-262 Ma, Menning, 1995) (Fig. 5.2), which has a thickness of 300 to 400 m in the studied wells. It overlies Lower Rotliegend volcanics at the base. The present lithotypes as characterized on resistivity image logs are divided into dune, sandsheet and alluvial fan deposits.

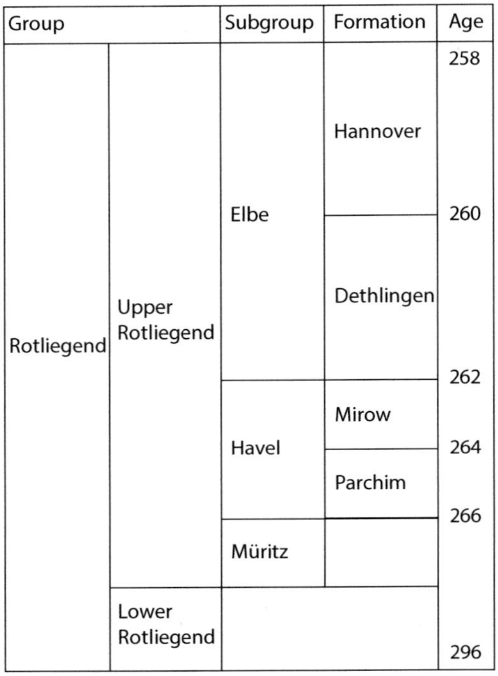

Figure 5.2: Stratigraphic subdivision of the Rotliegend in N-Germany (after Schöner (2006), stratigraphic ages from Menning (1995))

5.4 Materials and methods

The database comprises resistivity image logs, gamma ray logs and lithological logs of four wells including core material. The available samples were 14 plugs from core material for permeability measurements and 32 petrographic thin sections of four wells. Overall, plugs were taken from dune and sandsheet deposits at a reservoir depth between 4768 and 5251 m depth.

Plugs of 30 mm in diameter and 40 mm in length were used for permeability measurements, conducted in a high pressure flow cell at well-defined pressure and temperature conditions (p_{conf} = 6.5 and 32 MPa, p_{fluid} = 0.1-2 MPa, T = 25°C). In order to prevent by-pass of the permeating fluid, the samples were tightly sealed with a rubber sleeve before being placed in the water-filled autoclave. Argon was used as permeant for the pressure pulse decay measurements. Fluid pressures on both sides of the sample were continuously monitored, with the upstream pressure (p_1) held constantly at 1.8 MPa. The downstream pressure (p_2) within a previously calibrated reservoir ($V_2 = 1.02x10^{-05}m^3$) was allowed to equilibrate, starting from atmospheric conditions (Freund and Nover, 1995; Fowler et al., 2005). Apparent gas permeability coefficients were derived numerically by least-square fitting to the measured pressure decay data. Calculation was based on Darcy's law for compressible media (e.g. Hildenbrand et al., 2002):

$$\frac{dn}{dt} = \frac{V_2 \bullet dp_2}{dt} = -\frac{k_{gas} \bullet A(p_2^2 - p_1^2)}{\eta \bullet 2 \bullet x} \tag{5.1}$$

Here, the amount of mass (*dn,* mol) moving through the sample is calculated from the pressure change with time (dp_2/dt, Pa/s) in reservoir V_2. A [m^2] is the cross sectional area of the plug, η the dynamic viscosity [Pa·s], x the sample length (m) and k_{gas} [m^2] is the apparent gas permeability. A reliable Klinkenberg correction could only be performed for 7 samples. In order to obtain values for all samples, we additionally calculated Klinkenberg corrected permeability values (k_∞) according to a trend given by Jones and Owens (1980), which was based on more than 100 tight sandstones samples. The trend indicates larger slip-factors for rocks with lower permeability. Both methods yielded largely consistent k_∞-values. In general, Klinkenberg corrected permeability coefficients are 12% (±8%) lower than measured apparent gas permeabilities.

However, in the following we will concentrate on the interpretation of the apparent gas permeabilities. This is to avoid the usage of (likely inaccurate) extrapolation. As all measurements were conducted under the same experimental conditions, with the mean gas pressure at 1MPa, k_{gas}-values can be used for heterogeneity analysis (comparison between samples) of the present sample set.

All permeability experiments were performed parallel to bedding. To investigate the influence of deformation bands on bedding-parallel apparent permeability, different pairs of sample plugs were prepared (undisturbed host rock vs. samples with cataclastic deformation bands). The corresponding sample pairs were taken directly adjacent to one another from core material.

Additionally, company data on Klinkenberg-corrected permeability are provided for low confining pressures (1.8 MPa) using air as the permeant. Plugs parallel to as well as subnormal to bedding allow to access the impact of bedding horizons on flow. To reduce the influence of possible microfractures which formed as a response to external stress release after coring (Holt, 1994), permeabilities of plugs were measured at increasing confining pressures up to 32 MPa.

Based on this comprehensive dataset, transmissibility multipliers were calculated according to Manzocchi et al. (1999) for use in reservoir models:

$$T_M = \left[1 + \frac{t_{def}}{L}\left(\frac{k_{host} - k_{def}}{k_{def}}\right)\right]^{-1} \tag{5.2}$$

with T_M – Transmissibility multiplier, t_{def} – cumulative thickness of deformation bands in the calculated interval [m], L – Length of the calculated interval [m], k_{host} – the apparent host rock permeability [m^2], k_{def} – the apparent deformation band permeability [m^2]. For all calculations and assessments of deformation band occurrence around faults a damage zone width of 20 m has been assumed (Guo et al., 2009). This width was also adapted to assess transmissibility multipliers of deformation bands in a damage zone.

Petrographic methods include standard transmitted light microscopy, as well as cathodoluminescence microscopy (CL) performed with a HC3-LM, operated at an acceleration voltage of 13.5 kV and 0.6 - 0.8 mA/mm^2 beam current density. Of the 32 petrographic thin sections five were prepared without a cover slip for cathodoluminescence which is the only tool to outline quartz cemented cataclastic quartz grains. Optical porosity was measured from images of stained thin section and the representative elementary area (REA) was established by incrementally increasing the evaluated area on a photomosaic from 10x10 pixels up to the maximum area of the photomosaic (max. 4000*4000 pixels on a 12000*4000 pixel photomosaic within one layer). The REA was determined to be reached when the measured optical porosity stopped changing. This method was applied to samples stained with a low viscosity fluorescent dye. For the correlation of grain size and deformation band width, the average grain size was determined by averaging grain sizes measured along a grid in the host rock on either side of the deformation band. These data were taken adjacent to the site where deformation band width was measured. Orientations of deformation bands were plotted on lower hemisphere Schmidt nets.

The analysis and data mining of revised WellCAD data (resistivity image log, gamma ray log and lithological log) provided by the company was performed with the graphical user interface *rattle* (Williams, 2011) which is based on *R*, an environment and programming language for statistical computing. The *latticist* plugin allows the study of changes in orientation with increasing depth on marginal plots, which show a distribution for defined intervals (margins). The depth resolution of orientation data (dip direction, dip), derived from resistivity image logs embedded in WellCAD, can be adjusted by the number of created margins for the target variable depth. For this study the number of margins was set to four to allow a significant number of deformation bands per interval (≥ 40).

5.5 Results

The studied cataclastic deformation bands from core material all show a simple shear displacement between 0.4 mm to 15 mm. All samples contain discrete variations in grain size, be it in the form of foresets of grainfall or grainflow laminae in dune sandstones or deflation layers in sheetsand deposits.

5.5.1 Petrography

Most samples are similar in mineralogical composition and consist mostly of quartz (40 – 55%) with alkali feldspar and plagioclase (6 – 18% combined), lithoclasts (12 – 23 %) and iron oxide grains. Authigenic cements include quartz (0-15%), calcite (0-6%) and feldspar (0-5%). Accessory phases are always hematite and illite coatings and pore lining chlorite.

Petrographic analyses show that almost all studied deformation bands contain cataclasis of different degrees and cementation (Fig. 5.3 a-d). Cataclasis in the studied samples is followed by subsequent quartz cementation. Thus, cataclasis is often invisible in transmitted light microscopy but well shown in hot cathodoluminescence (CL)-microscopy, which displays intragranular fractures of detrital quartz cemented by quartz of different luminescence (Fig. 5.4 c). Thicknesses of deformation bands vary and do not necessarily depend on the amount of displacement along the deformation band. Generally thinner bands (0.2 - 1 mm thickness) occur in tightly cemented, densely packed and generally finer grained host rocks. Thicker bands, up to 10 mm wide, occur in less cemented layers and rather coarse grained intervals (Fig. 5.3 c). A correlation between the average grain size in the host rock and deformation band width exists for the studied cataclastic deformation bands with a coefficient of determination of $R^2 = 0.74$ (Fig. 5.5 a). All studied samples show a wide spread from below 0.1 mm to 0.9 mm in grain diameters of the host rock, occasionally including grains as large as 1.1 mm. The skewness of the distribution, as well as the Trask sorting only poorly correlate with the width of cataclastic deformation bands (Fig. 5.5 b). The median grain size resulted in the best correlation with deformation band width, only differing slightly from the average grain size in most cases (Fig. 5.5 b). The grain size distributions in our samples is not as narrow as results by Cheung et al. (2012), who experimentally established a correlation of the grain size distribution for compaction band occurrence. No such consistent trend could be established from the width of the distribution to the width of the cataclastic deformation bands presented here.

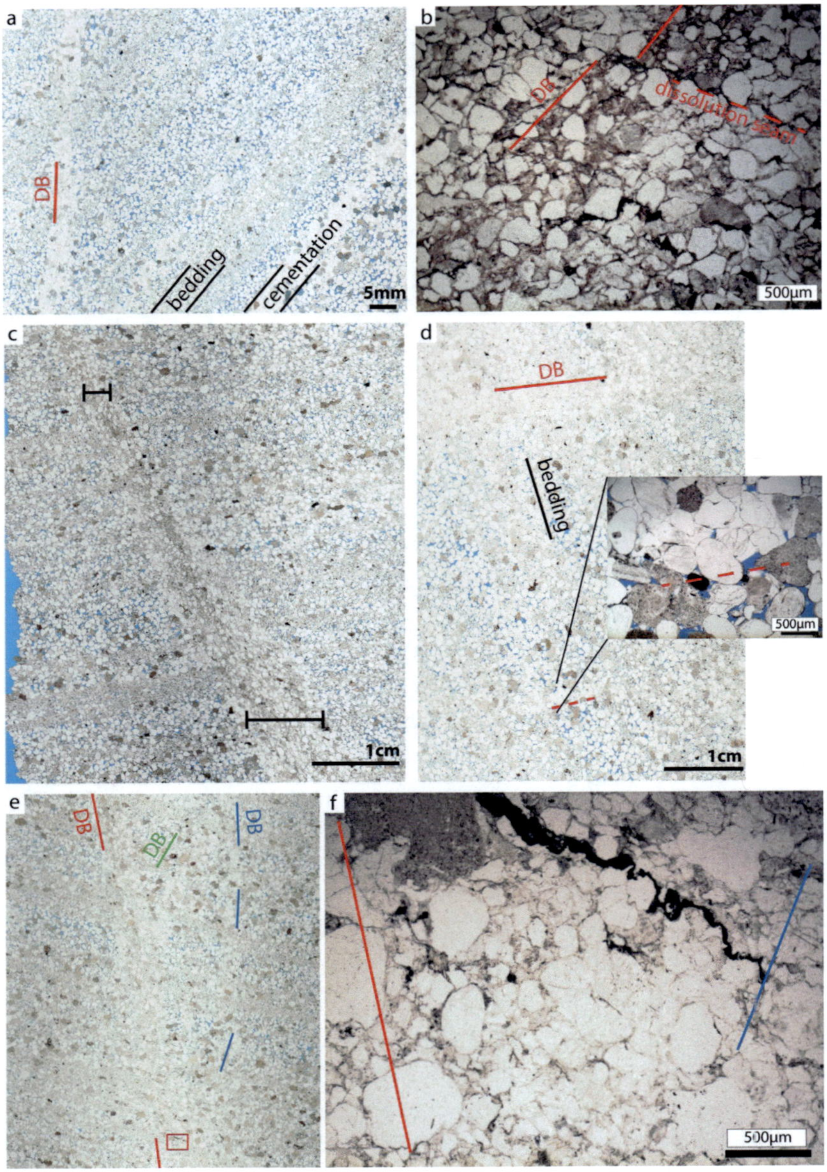

Figure 5.3 (previous page): a) Cataclastic deformation band at high angle to bedding marks a zone of low porosity, differences in grain size in the bedding and bedding parallel cementation. b) Compaction related dissolution seam parallel to bedding, which sets off a deformation band, indicates early formation of deformation band prior to chemical compaction. c) Cataclastic deformation band thickness (black lines) increases with an increase in average host rock porosity and grain size. d) Cemented area parallel to the deformation band (DB, red line) forming a load bearing structure in a relatively porous host rock. e) Thin section with three linking cataclastic deformation bands (DB), the red box covers the area of figure 5.3 f. f) In the concourse area of two linking cataclastic deformation bands an increase in pressure solution, but no increase in cataclasis is observed. Porosity colored in blue.

Incipient deformation bands show grain to grain contacts with minor pressure solution along quartz grain boundaries and quartz cements in adjacent pore space. They are oriented oblique to bedding, forming a load bearing framework. Incipient deformation bands are occasionally aligned in linear arrays at a high angle to bedding (Fig. 5.3 d). These arrays of grain-grain contacts and linear quartz overgrowth cements are also observed in the transition zone between the deformation band core and the undisturbed host rock (Fig. 5.4 a). In areas of two cataclastic bands merging into one band, no increase in cataclasis is observed, but only an increase in pressure solution (Fig. 5.3 e and f). Localized bedding-parallel cementation occurs predominantly in fine grained layers (e.g. Fig. 5.3 e).

Next to the 29 cataclastic deformation bands, only a few other types of deformation bands were observed which include two cementation bands (no shear component, continuous cementation with calcite at a medium to high angle to bedding, Fig. 5.2 a), one disaggregation bands (characterized by simple shear, disaggregation and dissolution) and one vein (characterized by simple shear, dilation and calcite cementation).

Petrographic differences of the two main lithotypes are restricted to textural differences, while no major differences were recorded from the mineralogical composition of given samples. The clay content increases to a maximum of 1% of the bulk volume in sand sheets.

The visible porosity inside the band was optically assessed by fluorescent staining and compared to the host rock (Fig. 5.6). Porosity inside a deformation band was analyzed in one sample by digital image analysis to be 2.1 ± 0.2 %. Host rock porosity ranges from 10.8 ± 0.7 % in the fine grained layers to 15.5 ± 0.5 % in coarser grained layers of that same sample. In another sample the porosity inside the deformation band was much lower at 0.18 ± 0.05 %. Host rock values in this sample range from 1.2 ± 0.08 % and 2.9 ± 0.3 % for two coarse grained bedding intervals of the host rock and 0.4 % to 1.3 % for one fine grained layer.

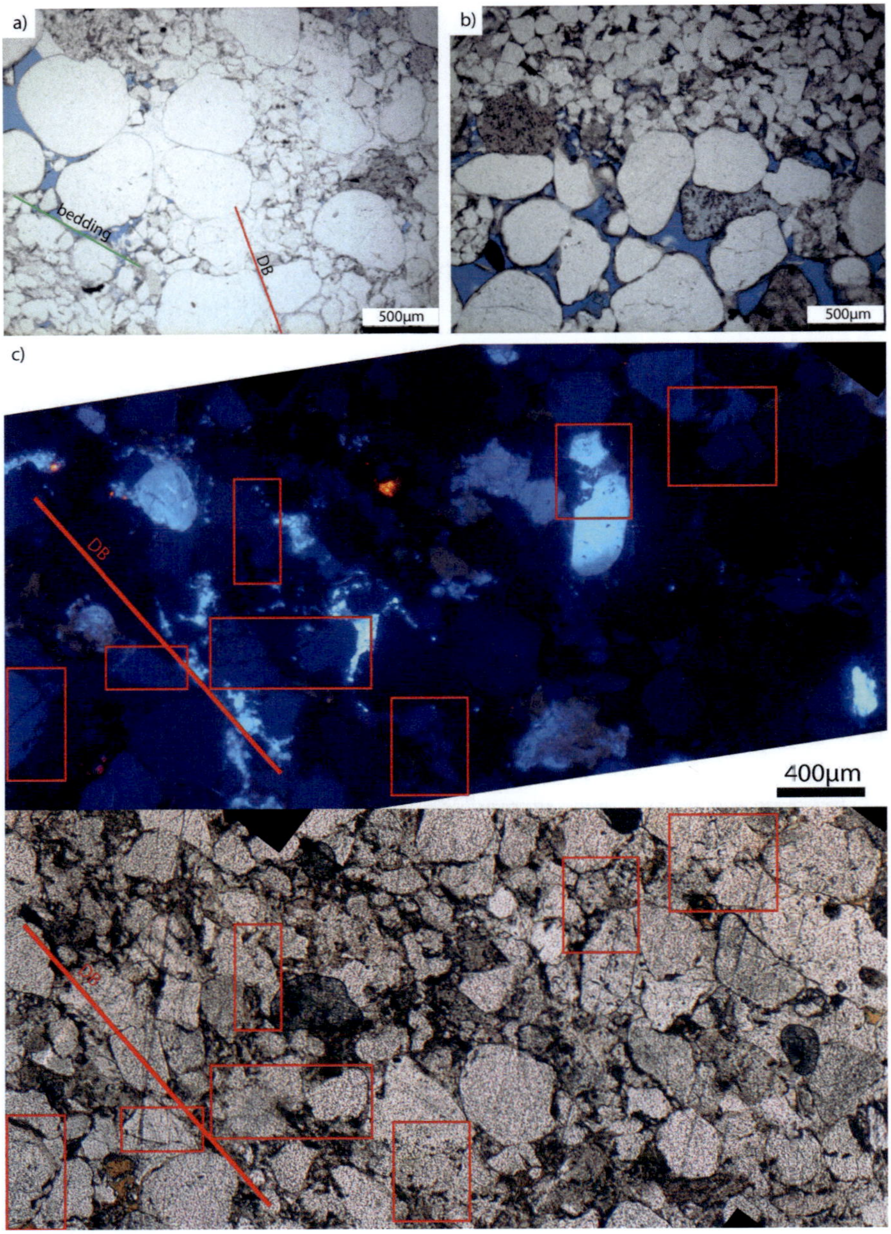

Figure 5.4 (previous page): a) Cemented margin of a cataclastic deformation band interpreted to be the remains of an incipient cataclastic bands. The pore space (blue color) in the rest of the thin section is uncemented (upper left and right corner). b) Mainly chloritic grain coatings (light green color) on coarse grained quartz grains preserve porosity in deeply buried sandstone samples. c) Comparison of a cataclastic band in cathodoluminescence (CL, top image, bright blue colors are illite and feldspar, dark blue colors quartz, orange colors are carbonates) and in transmitted light (bottom, ppl). Occasionally one can infer the cataclasis in ppl-images, but often the broken and healed grains of different quartz generations are only visible in CL microscopy.

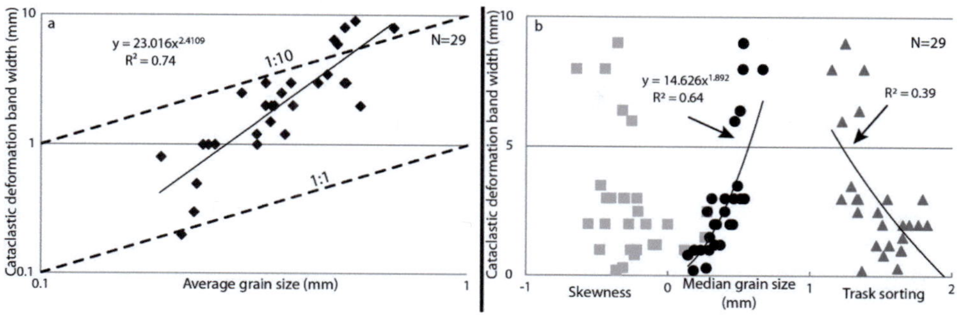

Figure 5.5: a) Correlation of average grain size in the undisturbed host rock versus deformation band width. The trend indicates that a larger grain size of the host rock results in wider deformation bands. Dashed lines indicate a 1:1 and 1:10 scaling relation. B) Correlation of skewness of the grain size distribution, median grain size in mm and sorting versus the width of cataclastic deformation bands. Except for the median grain size, the correlations are very poor.

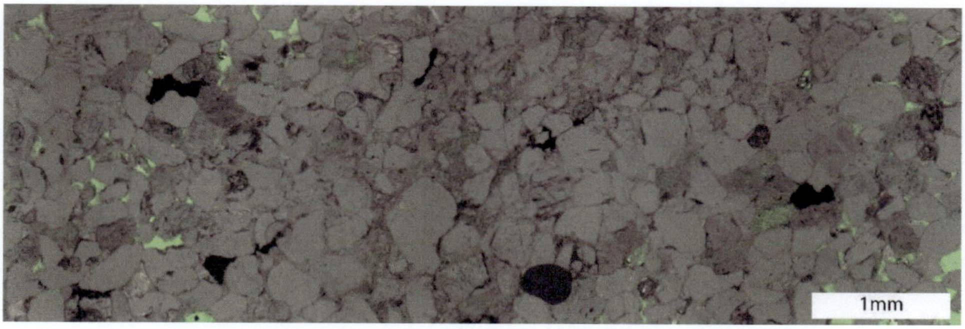

Figure 5.6: Thin section stained with fluorescent dye to highlight porosity (green color) in comminuted cataclastic deformation band. Dyed porosity inside the deformation band (central part of image) can clearly be differentiated from the porosity of the host rock (left and right side of image). Digital image interpretation and thresholding of the green fluorescent dye resulted in 2.1±0.2% visible porosity in the deformation band.

The two dissolution seams in the studied samples are enriched in illitic clay minerals, which are present as grain coating clays in most parts of the undisturbed samples. They developed parallel to bedding planes and foresets in dune layers, which suggests formation during chemical compaction. However, pressure solution postdates at least one set of cataclastic deformation bands, as derived from cross-cutting relations (Fig. 5.3 b). In most samples dissolution seams are concentrated in the fine grained intervals. In case of non-densely packed finer grained layers the area surrounding the dissolution seam is intensely cemented.

5.5.2 Data mining

Resistivity image logs were used to analyze the orientation and depth of the flow barriers. The conductive medium in the horizons is mainly the pore water. Flow barriers on resistivity image logs appear as more resistive features due to the loss or lack of porosity and are, in case of cataclastic deformation bands, oriented oblique to bedding and show a slight offset of the bedding planes. Bedding parallel compaction will most likely be masked by the bimodal sorting of the sediment. The lithotype association was applied consistently throughout the four different wells.

5.5.2.1 The dependency of deformation band density on lithotypes

The pre-picked lithologies in lithological log files were analyzed concerning the number of deformation bands per meter (DB/m). No consistent trends concerning the different lithological units (dunes, sand sheets, alluvial fans) could be established (Fig. 5.7). However, several field studies clearly demonstrate the accumulation of deformation bands in more competent beds (Schultz and Fossen, 2002).

Additional factors affecting deformation band density are the proximity to faults (Antonellini and Aydin, 1995), which are shown to be consistent if the host rock grain size at a similar distance to a fault is the same (Griffiths et al., 2016). The observations by Cheung et al. (2012) also point towards a grain size distribution-control on the occurrence of compaction bands which might also be applicable to deformation bands with a simple shear component. The number and displacements of observed faults derived from resistivity images cannot be linked to different lithotypes due to 1D data in wells.

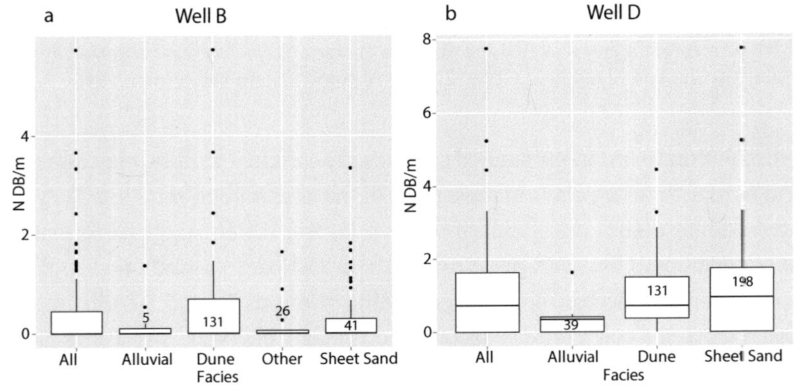

Figure 5.7: Box-whisker plots for two wells B (a) and D (b) of number of deformation bands per meter (DB/m) versus lithotype. Numbers next to boxes indicate the number of deformation bands analyzed in this lithotype. No clear accumulation of deformation bands in one lithotype as characterized from well logs and resistivity image logs can be determined. The dependency of deformation bands to fault throw, grain size and distance to faults had to be neglected due to the restricted 1D database.

5.5.2.2 The dependency of deformation band density on faults

Fault dependency was studied qualitatively from resistivity image logs and by calculating correlation coefficients of fault displacements versus number of deformation bands in the vicinity around faults. The lithotype dependency has been neglected due to the minor mineralogical differences of often closely spaced lithotypes. The maximum of observed displacements on faults was 30 cm. Published data from field exposures clearly show an increased number of deformation bands around faults and a correlation with fault displacement (e.g. Antonellini and Aydin, 1995; Rotevatn and Fossen, 2011; Griffiths et al., 2016). Since a well represents an arbitrarily oriented scanline across faults and surrounding lithotypes in the reservoir and lithotypes are similar, a correlation might be expected. However, well log data show no clear or consistent accumulation of deformation bands neither to their occurrence nor the displacement of small scale faults interpreted from resistivity image logs (Fig. 5.8).

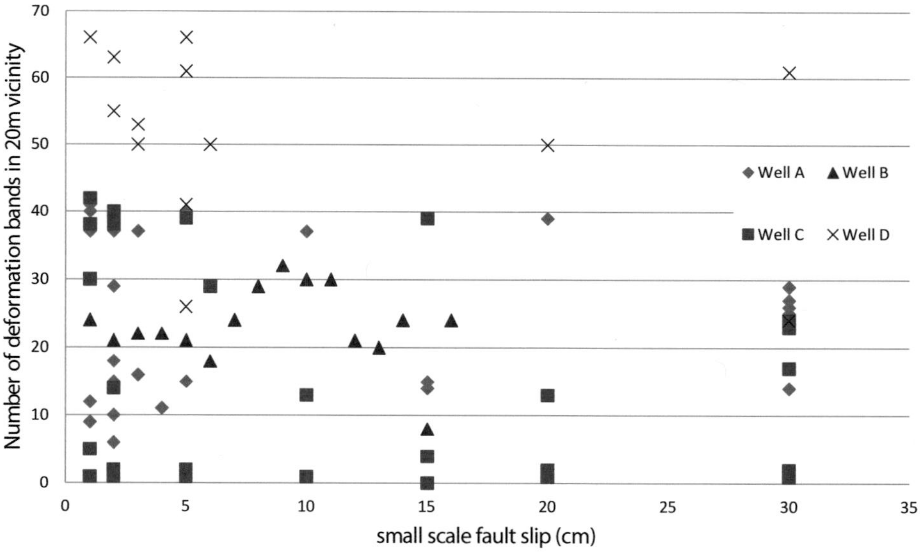

Figure 5.8: Plot of small-scale fault slip taken from image logs versus the number of deformation bands in 20m vicinity around the fault. No clear trend concerning the dependency of deformation band density in a certain interval versus fault slip can be established from the 1D image logs. A possible lithotype dependency has been neglected.

5.5.3 Marginal plots – Orientation data

The depth-dependent orientation trends were visualized by marginal plots, which split the dataset into four depth intervals (margins). Each depth interval in one well contains the same number of deformation bands. The intervals therefore can cover different lengths of the well section. The plots for the four studied wells are presented in figures 5.9 and 5.10. The first row in each plot visualizes the distribution of deformation bands along the well log interval. The second row shows the dip direction and number of deformation bands in each depth interval. The third row shows the dip and number of deformation bands in each depth interval with the according dip angle.

In well B and D, the deformation bands are accumulated in the second and third depth interval (blue and green line) (Fig. 5.9 a, b). These depth intervals cover parts of the Rotliegend stratigraphy of the Havel subgroup in both wells. The dip directions of resistive features in well B show a rather consistent bimodal distribution with a minor maximum at ~270° for all depth intervals and a second major maximum at ~90°. This reflects the consistent conjugate alignment of deformation bands along the well. The lowermost interval in well B (red line) comprises resistive features (veins) and conductive features (fractures) picked in the underlying Rotliegend volcanics interval of the log section, which explains the deviation from the consistent trend. The very minor maximum of dip directions towards ~ 180° in the third depth interval (green line) points to the accumulation of a few resistive features deviating from the consistent trend. The unimodal distribution of dip values ranges around 60° and reflects a conjugate extensional system.

Well D also shows a bimodal distribution of deformation band dip directions (Fig. 5.9 b). However, their strike does not match with those of well B, indicating a different local stress regime around well D during deformation band formation. Maxima at ~40° and 210° dip direction suggest a conjugate arrangement in the uppermost (purple line) depth interval. Dip directions also suggest a clockwise rotation of ~20° with depth, as indicated by aligning the peaks of the following two deeper depth intervals. At greater depth (blue to red line intervals) deformation bands dipping towards 300° become increasingly more prominent which indicates a similar clockwise rotation in strike. The dip values show an accumulation of shallower dips in the interval between 5079-5150 m (green line). In contrast to well B, dips are dominantly bimodal with one maximum at a more shallow dip of 50° and a second maximum at a steeper dip. There appears to be an increasingly steeper dip with increasing depth from ~70° to 80°. The change in strike and steepening with depth may indicate the influence of a flower structure (Woodcock and Schubert, 1994).

In well A, the number of deformation bands increases with depth (Fig. 5.10 a). Dip directions of deformation bands show a bimodal distribution with maxima at ~90° and 270°, indicating a conjugate set of deformation bands. Similar to well D, steepest dips of ~80°

are observed at the deepest depth intervals. An anticlockwise rotation in dip direction is indicated by arranging the peaks of the top three intervals (purple, blue and green line) around 90°. A second more prominent azimuth of ~200° is present in the lowermost depth interval. The lowermost interval (5199-5424m) for this well only comprises fractures of the Rotliegend volcanics causing the deviation from the consistent trend.

In well C the deformation bands are accumulated in the second and third depth interval (blue and green line) (Fig. 5.10 b). The distribution of dip directions has four rather consistent maxima at ~290°. The second maximum is located between 70-120°. The 70° dip direction corresponds well with the dip direction extracted from seismic line interpretation and is present in the lowermost interval a well as subordinately in the shallowest interval (red and purple lines). The dominant dip values steepen with depth (blue to red line intervals) from ~55° to 70°. However, steepest dips of 80° are also dominant in the shallowest depth interval (4888-5073 m, purple line).

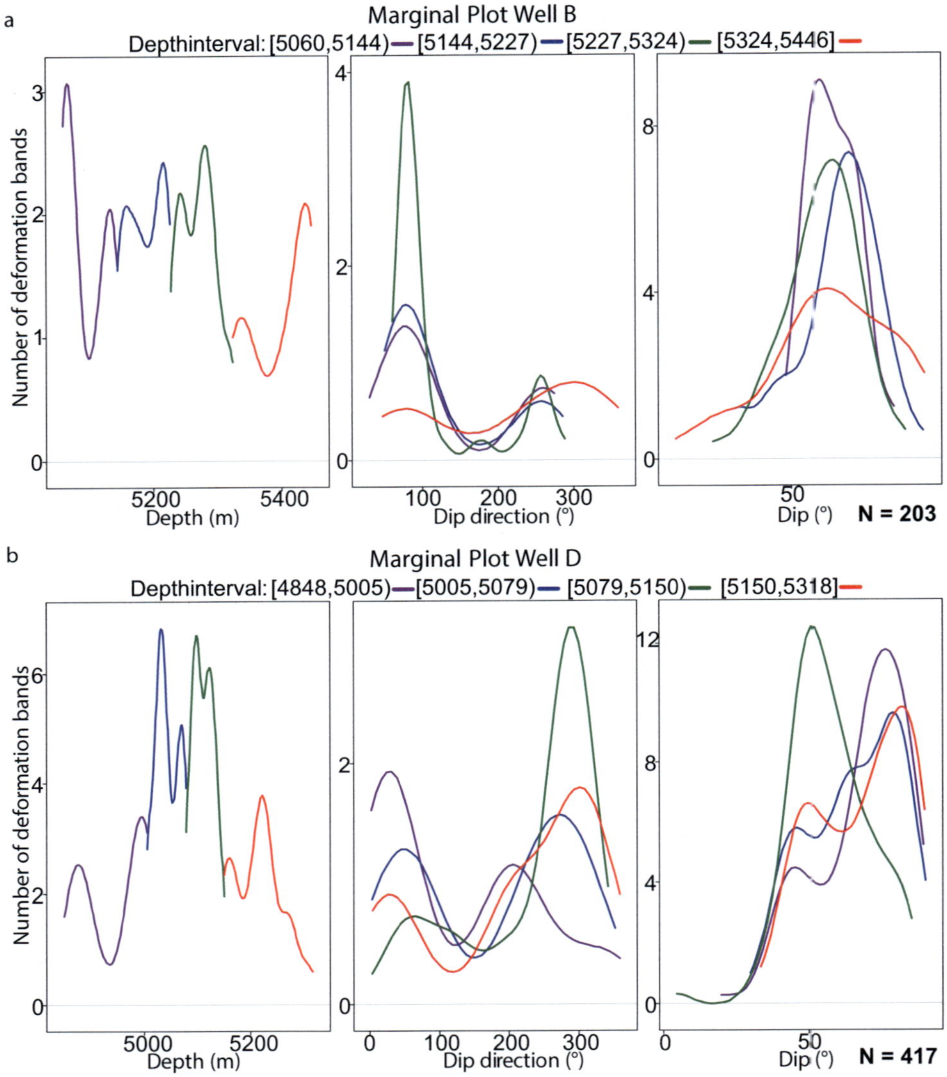

Figure 5.9: Marginal plots for wells B (a) and D (b) for different depth intervals. The total number of deformation bands is given as N. The four depth intervals (margins) from top to bottom are a purple, blue, green and a red line. The according depth intervals are given at the top of each plot. First row: Distribution of deformation bands along the log interval. Second row: Azimuth of deformation bands in each depth interval, note the accumulation in well D between 5079-5150m (green interval). Third row: Dip of the deformation bands in each interval; note the inhomogeneous distribution in well D in comparison to well B.

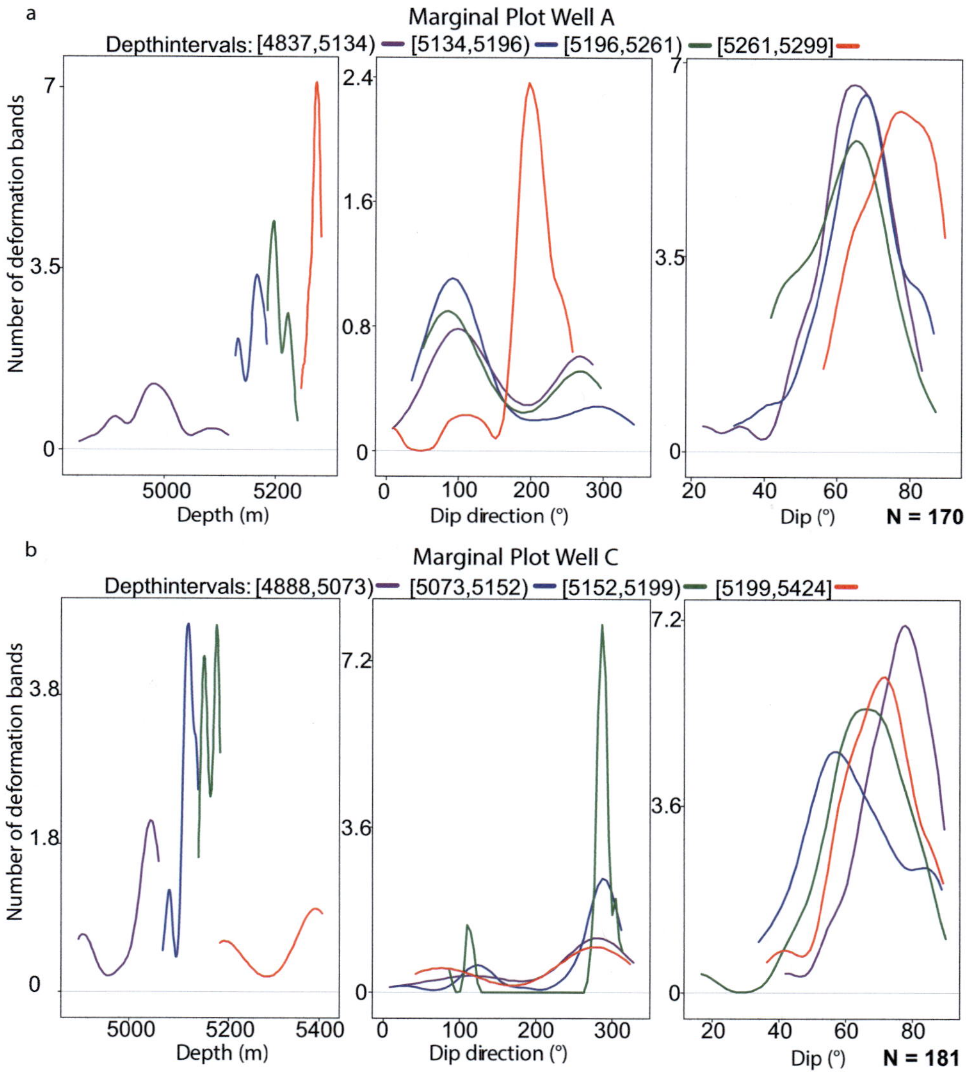

Figure 5.10: Marginal plots for wells A (a) and C (b) for four different depth intervals. First row: Distribution of deformation bands along the log interval. Second row: Azimuth of deformation bands in each interval. Third row: Dip of deformation bands in each interval. Well C: lowermost interval (5199-5424) represents fractures and veins in the volcanic units.

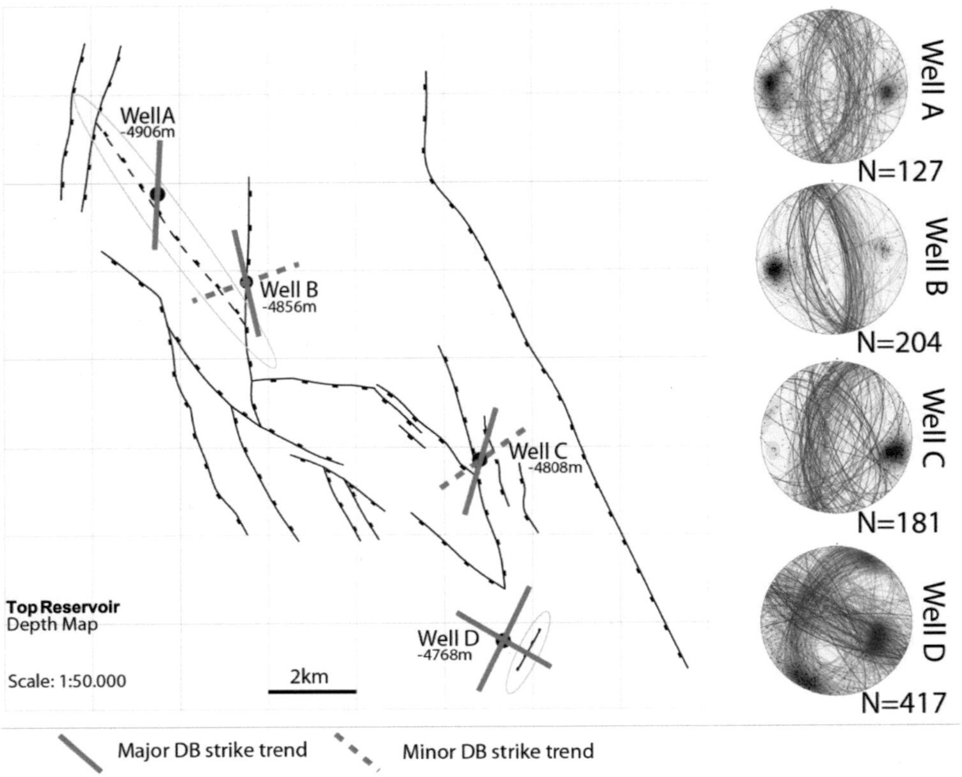

Figure 5.11: Structural map of the top of the reservoir. Black lines are normal faults, grey lines on the well locations represent the strike of small scale flow barriers as interpreted on resistivity image logs, dashed lines represent minor strike sets deduced from resistivity image logs. Circles indicate areas which might be reassessed by small scale structural data with a depth resolution. The denoted depths represent the depth of the top of the reservoir unit. Only resistive features (deformation bands) oblique to bedding are visualized in Schmidt net plots.

Given that deformation bands form sub-parallel and prior to faults, deformation bands and their orientation can be used to infer the orientation of larger scale faults. Thus, faults depicted on seismic lines and inferred from production data may be validated by faults and deformation bands from the resistivity image logs and vice versa. In case of well A, the resistivity image log, covering all of the Havel subgroup, shows no small scale resistive features indicating faults of the given strike (145°) as is noted in the structural map (Fig. 5.1). The fault which is noted on the structural map was derived only from production data (Fig. 5.11, ellipse around well A). This fault strike derived from production data can only be seen in small scale faults, fractures and resistive features (veins) of the underlying Rotliegend volcanics. However, it is not reflected by resistivity anisotropies (deformation bands, faults, veins) in the reservoir unit and must thus be questioned. In case of well D, the resistivity log data (strike of deformation bands, faults, veins) suggests that only one orientation of faults interpreted from seismic sections matches one of the two main orientations of small scale resistive features. Additionally, the second orientation is accumulated in a certain depth interval (green line Fig. 5.9 b). This may indicate a subseismic fault trend (Fig. 5.11, circle at well D).

5.5.4 Permeability

The apparent permeability at high confining pressures decreases nonlinearly, which is related to the closure of microfractures. Measured data were fitted by exponential regressions, which are given in Figure 5.12 a-d. Generally, the decrease from 6.5 to 32 MPa confining pressure was less than 1 order of magnitude (Fig. 5.12 a-d).

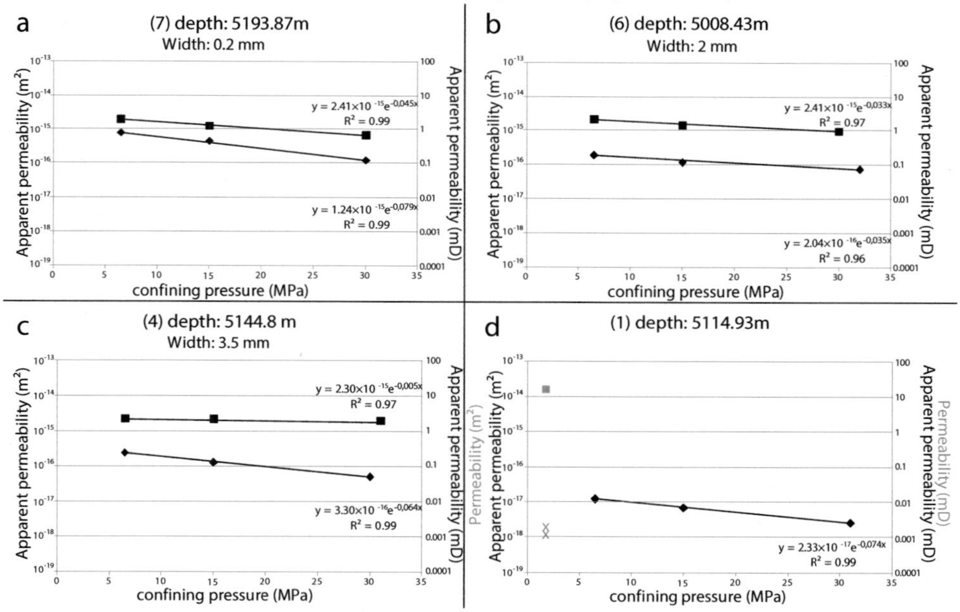

Figure 5.12: Apparent permeability results for four pairs of samples (1,4,6 and 7). Diamonds are measurements parallel to bedding of samples including a deformation band, boxes are samples without a deformation band parallel to bedding and crosses are measurements perpendicular to bedding. Apparent permeability decreases nonlinearly and the main reduction occurs at confining pressures below 6.5 MPa. Thin deformation bands (sample 7) develop less of a apparent permeability anisotropy than wider deformation bands (sample 4). The decrease in apparent permeability correlates with the deformation band width in the three samples where the correlation was possible. Plot for sample 1 visualizes the apparent permeability anisotropy of the regular host rock with samples measures perpendicular to bedding (crosses, only in d) as opposed to measurements parallel to bedding (box, only in d). The displayed samples were collected within 5 cm of the sampling location of sample 1. The grey error bars show the impact of the Klinkenberg correction. Apparent permeability is higher than the Klinkenberg corrected values.

The stress dependency of measured apparent permeability can well be described by an exponential function (cf. David et al., 1994). The stress dependency factor γ of our samples ranges between 0.031 and 0.074 MPa^{-1}, which are within or slightly above those given in literature for tight sandstones (Yale 1984, compiled in David et al., 1994).

The permeability of the host rock samples (Fig. 5.12, squares) are larger than those of adjacent samples containing a deformation band (Fig. 5.12, diamonds). At higher confining pressures this difference remains almost constant for each sample set with less than one to more than one order of magnitude (Fig. 5.12 a, b). Only sample set 4 shows a continuous increase in permeability reduction from one to nearly two orders of magnitude during increasing confining pressure (Fig. 5.12 c).

Permeability reduction of sample 7 (A-B) is smaller than in other more porous samples at 30 MPa confining pressure (Fig 5.12 a, squares), which is due to specific microstructual differences. Here, deformation bands are located in a fine grained, well cemented and densely packed host rock. The thin nature of the deformation band in this fine grained host rock only reduces the permeability by only 0.7 orders of magnitude with respect to the undisturbed sample at 30 MPa confining pressure (Fig. 5.12 a). Coarser grained host rocks favor the development of wider deformation bands, which result in a larger permeability reduction (Fig. 5.12 b, c).

The permeability across dissolution seams and bedding-parallel cementation could only be inferred from low confining pressure data provided by the company. These data indicate that the permeability measured normal to bedding is 1 to 4 orders of magnitude lower than corresponding permeability parallel to bedding (Fig. 5.12 d). This reduction corresponds with observed bedding parallel cementation and chemical compaction by either quartz or calcite and bimodal grain size distributions on a microscopic scale.

Two experiments with subsequent stress reduction (30 MPa to 6.5 MPa) on host rock samples (3B and 7B) resulted in lower apparent permeability of 10 % and 15 %, respectively, which could either indicate irreversibely closure of microfractures or the collapse of pore space at larger confining pressures.

During the measurements several samples fractured at low as well as at high confining pressures. One sample completely disintegrated to sand, others partly disaggregated. This suggests that samples are partly underconsolidated in the reservoir, implying early fluid charge and the presence of fluid overpressures.

Transmissibility multipliers were calculated for peak confining pressures of 30 MPa using the given apparent gas permeability data and calculated for the width of the deformation band. These transmissibilty multipliers (equation 5.2) were calculated for deformation band frequencies that were determined from the actual resistivity image logs (sample series

A), a worst case scenario (sample series B) was assumed for 100 DB/m and a frequency derived from a published field example (sample series C) from Johansen and Fossen (2008) (Tab. 5.1). Thresholds for the sealing efficiency are based on transmissibility multiplier calculations by Shipton et al. (2005), who suggest the abundance of sealing deformation bands for $T_M < 0.0005$ on geological timescales and $T_M < 0.001$ on production timescales. Transmissibility multipliers show that the sampled deformation bands do not completely seal off fluid flow in the reservoir under the given assumptions, not even at reservoir confining pressure conditions (Tab. 5.1). Contrary to this, reservoir data have shown that flow barriers do stop flow parallel to bedding.

Table 5.1: Compilation of transmissibility multiplier (T_M) calculations for samples 4, 5, 6 and 7 with different deformation band densities (DB/m). Input parameters are apparent permeability calculated for the deformation band width and thickness. The length of the reference interval is 20m. Confining pressures at which apparent permeability has been determined are given.

	Sample	k_{def}	k_{host}	P_{conf}	DB/m	Thickness (t)	T_M	Leak/Seal	
	No.	m²	m²	MPa	N	m	(-)	Geological timescale	Production timescale
A	4	4,53E-18	1,95E-15	30	1	3,50E-03	2,32E-02	leak	leak
	6	2,74E-18	2,40E-15	32	0,5	1,50E-03	5,20E-03	leak	leak
	7	9,08E-19	6,23E-16	30	0,5	2,00E-04	1,46E-03	leak	leak
B	4	4,53E-18	1,95E-15	30	100	3,50E-03	2,32E-03	leak	leak
	6	2,74E-18	2,40E-15	32	100	1,50E-03	5,20E-03	leak	leak
	7	9,08E-19	6,23E-16	30	100	2,00E-04	1,46E-03	leak	leak
C	4	4,53E-18	1,95E-15	30	17,25	3,50E-03	2,32E-03	leak	leak
	6	2,74E-18	2,40E-15	32	17,25	1,50E-03	5,20E-03	leak	leak
	7	9,08E-19	6,23E-16	30	17,25	2,00E-04	1,46E-03	leak	leak

5.6 Discussion

5.6.1 Petrography

The optical evaluation of deformation bands from thin sections, coupled with apparent permeability measurements better constrains the impact of deformation bands on fluid flow in the reservoir. The absence of grain coatings and the high reactivity of fractured quartz grains enhances the cementation of cataclastic grains (Fisher et al., 2000; Fossen et al., 2007). In the studied samples such masking of cataclasis in deformation bands is only visualized by cathodoluminescence microscopy, which is recommended to verify the presence of brittle deformation in grains.

A correlation of deformation band thickness and grain size was described by Wennberg et al. (2013, their Fig. 12 and references therein) for carbonates and sandstones. Our results for sandstones are in accordance with their findings with a good correlation of cataclastic deformation band thickness versus the host rock grain size adjacent to the band (Fig. 5.5). Such correlation also links to our petrophysical measurements, which show an increase in the apparent permeability reduction with increasing deformation band width. It allows a first apparent permeability prediction and thus points to the role of deformation band width on fluid flow in a reservoir (Fig. 5.13). A higher number of data should be generated to further support the overall trend.

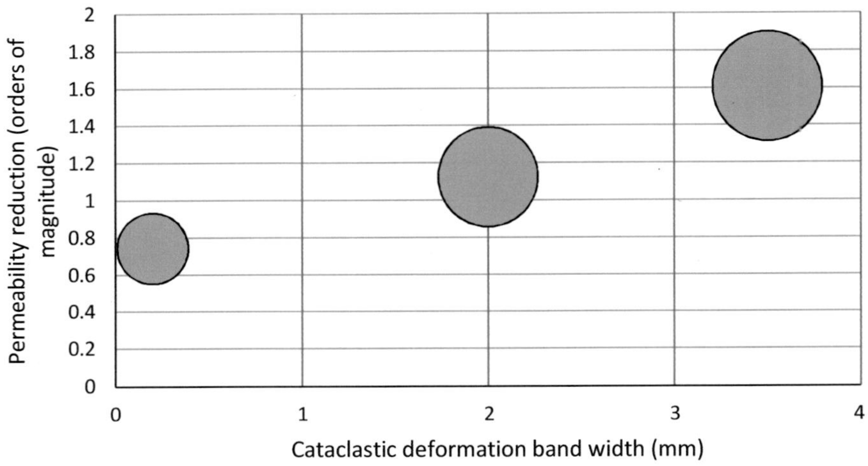

Figure 5.13: Apparent permeability across cataclastic deformation bands inferred from deformation band width, based on a combination of grain size vs. cataclastic deformation band width correlation plot and cataclastic deformation band width and apparent permeability reduction plot. The point size corresponds with the grain size. This correlation enables to predict the reservoir quality if deformation bands are to be expected in a reservoir unit.

The experimental observations of Cheung et al. (2012) demonstrate the important role of grain size distribution on the physical properties during the formation of a deformation band. In their experimental setup the authors created compaction bands, which are end members of deformation bands without any considerable shearing. Sandstones which have a comparably narrow grain size distribution (spread below 300 μm) formed localized compaction bands, whereas comparably wide distributions (spread of 700 μm) resulted in distributed cataclastic flow (Cheung et al., 2012). Our localized cataclastic deformation bands, however, formed in a host with a wide spread of grain size distributions (>300 μm) and with shear displacement. Thus, experimental results cannot be transferred to our data with respect to localization. Furthermore, no correlation is evident for the deformation band width versus the skewness of grain size distributions and the sorting of the host rock (Fig. 5.5 b).

The evolution of deformation bands can well be explained by localization of stress at load-bearing frameworks which form during initial mechanical and chemical compaction. Small cemented arrays at high angle to bedding and crossing porous beds may represent such load-bearing framework with solution along grain boundaries and precipitation in the adjacent pore space (see Fig. 5.2 d, Fig. 5.14). These frameworks may evolve towards a discrete cataclastic deformation bands (Fig. 5.14) due to the overall cementation and strain hardening (Antonellini and Aydin, 1995; Fossen, 2010; Ballas et al., 2012; Soliva et al., 2013).

The bedding parallel cementation is mainly concentrated in the finer grained parts of the sandstones, although locally coarser grained layers are also affected (Fig. 5.3 a, d, e). Samples point to a smaller coat coverage in fine grained beds causing higher reactive quartz surfaces, resulting in higher cementation (Bloch et al., 2002).

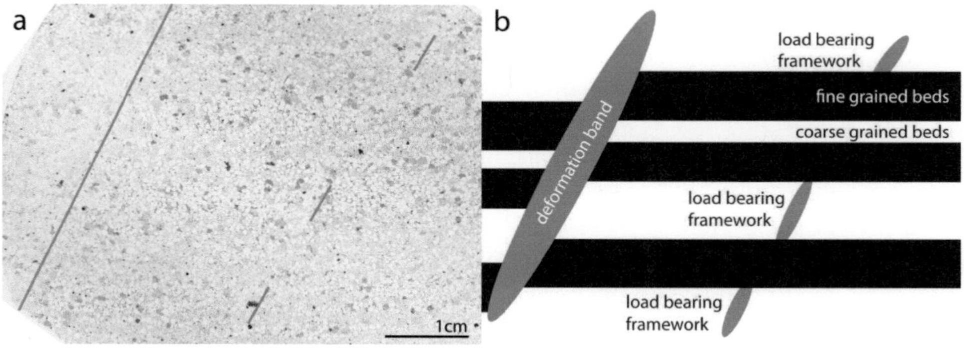

Figure 5.14: a) Thin section of a cataclastic deformation band (grey line) and load bearing structures (dashed line) in the high porosity beds aligned parallel to the deformation band. The image on the right is the same as in Figure 5.2 d. b) Conceptual model of the formation of cataclastic deformation bands in high porosity sandstones. The formation of a load bearing framework during chemical compaction by cementation and/or pressure dissolution acts as a precursor from which a cataclastic deformation band develops. The load bearing framework is developed sub-parallel to deformation bands which supports the hypothesis. Porosity is higher in the coarse grained intervals.

5.6.2 Data mining

Datamining, especially marginal plots, proved to be a good method to add a depth resolution to spatial orientation data. The visualization of changes in dip direction and dip with increasing depth cannot be made on one Schmidt net plot. The rotation of strike with increasing depth and the steepening of deformation band dip angles possibly resembles a flower structure (well D), although the geometry of the fault planes cannot be visualized with the limited dataset provided for this work. The local variations in strike and dip of the deformation bands are ascribed to changes in the local stress field and can be compared to the interaction of sub-seismically observable faults with seismically observable faults (cf. Johansen et al., 2005, their Fig. 4). The deviation of deformation band strike from seismically observable fault strike in well C is attributed to interactions between closely spaced normal faults and similar to the setting presented in Antonellini and Aydin (1995, Fig. 7).

The resistivity image logs of well A show an absence of sub-seismically observable faults, deformation bands and veins matching the strike of a large seismically observable fault on the structural reservoir map. However, several field studies show the occurrence of deformation bands around normal faults (e.g. Antonellini and Aydin, 1995; Johansen and Fossen, 2008; Fossen and Rotevatn, 2012) and their use as a proxy of nearby faults. Such a proxy may thus verify reservoir scale faults interpreted from seismic sections (in our case subsalt) or production data. Although deformation bands generally predate fault formation (Antonellini and Aydin, 1995) and are thus not a sufficient criteria to infer faults, the mismatch of resistivity- and reservoir map data led to a successful revision of the structural map.

Dependencies of deformation band densities in different siliciclastic lithologies could not be established, which may be ascribed to similar lithotypes. Furthermore, additional sets of deformation bands related to additional faults in the vicinity of the well affect any regular distribution.

5.6.3 Permeability

Apparent permeability data of samples including a cataclastic deformation band and according undisturbed host rock samples show that the studied cataclastic deformation bands have a negative impact on production.

The maximum reduction of two orders of magnitude for a measured cataclastic deformation band is in the range of previously measured samples (e.g. Fossen and Bale, 2007; Tueckmantel et al., 2012). Experimentally created compaction bands within sandstones led to a permeability reduction by two to three orders of magnitude (Vajdova et al., 2004; Baud et al., 2012; Deng et al., 2015). Brittle shear faulting in an experimental setup by Zhu and Wong (1997) resulted in a permeability decrease of less than one to two orders of magnitude as opposed to the undisturbed rock.

While all our apparent permeability data were measured along bedding and lowers across sheared cataclastic deformation bands, the permeability reduction is higher across cemented bedding and bedding parallel compaction bands. This corresponds with results of Baud et al. (2012) who reported compaction bands parallel to bedding to form the most efficient flow barriers. The application to a field scale, however, needs to include the assessment of the connectivity of all deformation band types in three dimensions to form extensive barriers to fluid flow (e.g. Fossen and Bale, 2007).

The permeability of a sample containing a deformation band is mainly controlled by the low permeable deformation band. Considering the entire sample length results in larger permeability coefficients compared to the thin deformation band itself (cf. Main et al. 2000; Baud et al. 2012). Since the deformation bands will be the controlling factor in the permeability difference between samples with and without deformation bands (cf. Main et al., 2000; Baud et al., 2012), the permeability measurements were also evaluated regarding the width of the deformation band. Apparent permeabilities of the deformation band itself is one order (wide cataclastic deformation bands) to two orders (narrow cataclastic deformation bands) of magnitude lower than the bulk apparent permeability of such host rock sample containing the deformation band.

The assessment of the transmissibility multipliers, calculated with the apparent permeability coefficients of the deformation band, will only significantly change to be sealing, if the permeability anisotropy exceeds more than three orders of magnitude irrespective of the number of deformation bands per interval.

Relative permeability reduction after stress release was attributed to irreversible granular compaction resulting in permanent deformation (cf. Saillet and Wibberley, 2013). Although no repeated stress cycles were performed, the permeability-stress hysteresis is highly probable. If a less confined host rock sample after peak confining pressures results in lower permeabilities than the same sample at high confining pressures, permeability stress hysteresis as measured in other siliciclastic rocks (e.g. Faulkner and Rutter, 2000) can strongly be inferred.

As several plug samples were (partly) disintegrated after application of confining pressures of up to 32 MPa we suggest that the critical pressure marking the onset of grain crushing and pore collapse (e.g. Wong et al., 1997; Tembe et al., 2008) has been exceeded during the measurements. We suggest that reservoir pore pressures were very high to maintain host rock stability at current burial depth. Thus, significant effective stress increase could lead to depletion induced compaction failures (cf. Schutjens et al., 2004). Maintaining the initial pore pressures might be essential to prevent reservoir integrity failures.

Orientation dependent permeability anisotropies, parallel and vertical to bedding can be attributed to pore shape anisotropies, grain size differences (clay size to sand size), sorting, sedimentary fabric, compaction, cementation, microcracking and pressure solution (Benson et al., 2005; Louis et al., 2005; Armitage et al., 2011). Sedimentary fabric induced permeability anisotropies of sandstones of up to 4 orders of magnitude are in accordance with findings of Armitage et al. (2011).

5.7 Conclusions

Almost all deformation bands are cataclastic, the fractures cemented by quartz. Cemented cataclastic fractures are not visible in transmitted light microscopy but cathodoluminescence microscopy. Intersections of dissolution seams and deformation bands indicate that deformation bands formed prior to and during chemical compaction.

Marginal plots of resistivity logs highlight the rotation of deformation bands and fractures with depth, which may point to a flower structures. Lithology dependent variations of deformation bands in the reservoir units could not be established from image logs and cores due to the similar mineralogy of the beds.

Data mining from image logs were successfully used to re-evaluate structural reservoir maps and seismic scale faults. It may further indicate subseismic fault trends at specific depth intervals.

A scaling relation for cataclastic deformation bands correlates an increase in deformation band width to an increase in average host rock grain size.

Well cemented and densely packed fine grained host rocks contain thinner deformation bands with a lower permeability contrast of less than one order of magnitude. More porous and coarser grained host rocks reach permeability contrasts of up to 2 orders of magnitude along bedding. However, the impact of bedding-parallel dissolution seams and cementation parallel to fine grained bedding, has a larger impact on flow by up to 4 orders of magnitude.

The failure of some samples at or below 30 MPa confining pressure may be indicative of formation intervals at high fluid pressures, which during reservoir depletion may lead to depletion-induced compaction failure, reducing flow.

6 Conclusions and outlook

6.1 Conclusions

Reservoir quality in the studied setting northeast of Hannover is largely controlled by differences in syntaxial quartz overgrowth cementation. In samples containing continuous, subparallel to disordered radial chlorite coatings occasionally completely replacing smectite, syntaxial quartz overgrowth cementation is lowest. Here, the most porosity is retained. In samples which lack these coatings and only contain pigmented hematite dust rims on the grain surfaces, quartz cementation is pervasive and intergranular porosity is largely occluded.

Burial model evaluation based on kinetic quartz and illite cementation modeling is a very good opportunity to control the quality in uncertain settings. Due to the different response to thermal exposure of the organic matter and quartz and illite cementation, a clear differentiation can be made which will favor one reconstruction over the other. Additionally, K/Ar-dating can validate simulation results of the formation of fibrous illite in the kinetic illite model. The formation of illite in the applied model largely relies on the availability of kaolinite and K-feldspar and once the precursors are consumed, the reaction ceases. The initiation of this reaction is kinetically controlled and thus, a timespan during which illite formed can be correlated to K/Ar-ages from the region. Kinetic quartz cementation models describe the formation of syntaxial quartz cements as a function of available surface area over time and temperature. Systematic mismatches between modeled and measured amounts of quartz cement volumes additionally indicate an inaccurate thermal reconstruction. However, the strong sensitivity and accuracy of the illite model is the key to this more accurate reconstruction, which matches well with other reconstructions of the region.

The improvement of the accuracy of burial models, especially using illite kinetic constraints, will impact the assessment of the timing of hydrocarbon generation.

The amounts of observed quartz cement volumes correlate well with the measured amounts of grain coat coverage and are expected to be responsible for the differences in cementation in the Penrith Sandstone.

The color alteration surrounding deformation bands at outcrops and diagenetic alterations within macroscopically different deformation bands can be used to relatively date the formation of these potential flow anisotropies. Information on enhanced permeability along the deformation bands is preserved by the bleaching phenomena confined to some deformation bands. The bleaching of hematite from grain surfaces is possible since illite coatings inhibited quartz overgrowth cementation while keeping the intermixed hematite in contact with the pore fluid. Some millimeter away from the deformation band, illite coatings are still stained red. Similarities in strike are used to determine the relative timing in regards to the Tertiary Armathwaite-Cleveland dyke. The pervasive cementation of another cataclastic deformation band generation at the outcrop indicates a formation prior to burial diagenetic quartz cementation. Where illite coatings are absent or only present covering small amounts of the grain surface, the lithology still contains red pigmented hematite dust rims, overgrown by quartz cement. This allows the relative dating of different fluid chemistries impacting diagenetic alterations in the formation.

State-of-the-art scanning electron microscopy (SEM) analysis using a FEI QEMSCAN© system results in quantitative mineralogical data and eases the detection of low-concentration detrital and authigenic phases in the studied sandstone samples. However, the QEMSCAN© methodology, using backscatter electron images and energy dispersive X-ray spectroscopy, lacks the possibility to accurately image and analyze thin mineral coatings on substrate grains without measuring the underlying mineral phase as well. This is an artifact of the excitation volume, a function of acceleration voltage, inherent to all scanning electron microscope analyses. Since the thickness of coatings is in the range of few µm, even down to a nanometer scale, the EDX signal acquired always contains information of the underlying or adjacent grain/mineral phase. This results in an incomplete measurement of grain coatings with this method, which will result in wrong assumptions of the actual coat coverage of the grain. Light-based petrographic analyses can perform this task with ease due to very characteristic interference colors of illite and hematite-stained illite in cross- and plane polarized light.

The delineation of a simple impact of the depositional environment on the presence or the extent of grain coatings in fluvio-eolian sandstones cannot be made on a Southern Permian Basin-wide sample series. Neither is a clear preference visible to more or better grain coated samples in a certain depositional environment, nor is there a clear preference to generally accumulate grain coatings in a certain depositional environment. Samples from four Rotliegend areas differ in source area, burial depth, cement phases, detrital composition, and diagenetic alterations. This might additionally explain the poor correlation. However, even on a small (km-) scale (e.g. NE-Hannover samples) the ranges of poor- and well-coated grains differ very little between the environments and so do the averages and means of these correlations.

Additional correlations with grain size, sorting or detrital minerals did not result in a universal trend applicable to all settings. Only locally the grain size correlates with the observed amounts and extent of grain coatings. The detailed petrographic analysis of each sample is therefore imperative to calibrate a diagenetic model when it comes to the relevant category of grain coat coverage. This investigation strongly confirms that the emplacement and presence of grain coatings is not only controlled by a single process but by a complex interaction of numerous parameters.

The impact of using universally applicable data is shown to overpredict quartz cement volumes in samples featuring lower universal grain coat coverages than the measured coat coverages, and to underpredict quartz cement volumes where average grain coat coverages are higher than the measured data. This will lead to a large spread around the 1:1 line of measured versus modeled quartz cement volumes during calibration. This cannot be easily accommodated for during model optimization, which will ultimately impact the predictive capabilities of the model. The mismatches of reservoir quality modeling on reservoir evaluation are too severe to determine universally applicable dependencies of grain coat coverages from this Rotliegend database.

The studied cataclastic deformation bands from northern central Germany only reveal the necessary microstructures needed for classification in cathodoluminescence. The interaction of chemical compaction and structural features allows the delineation of relative temporal relations. Deformation bands in this case formed prior to and concuccrently with chemical compaction.

The statistical evaluation and splitting of data into depth groups (marginal plots) result in a new approach to interpret deformation bands from core material. Scaling relations based on core material in cooperation with derived systematics in petrophysical measurements can be used to evaluate the impact of deformation bands on flow properties.

Core sample failure below 30 MPa confining pressure during petrophysical analysis may be indicative of formation intervals at high pore fluid pressures that, during depletion, might lead to induced compaction failures, and hence to reduced flow.

6.2 Outlook

During the work on this project, a field work campaign in active fluvio-eolian depositional environments in the United States was performed to identify grain coatings in these systems. Samples acquired in the Algodones Dunes, Imperial Valley, California, the Mesquite Flat Sand Dunes, Death Valley, California, and Coral Pink Sand Dunes, Utah, were collected from different depositional environments within the areas to generate a database to primarily detect and describe early diagenetic or depositional grain coatings. The processing and analysis of 60 samples is ongoing at the time of submission of the thesis but preliminary results indicate a confirmation of the independence of grain coatings and the depositional environments. Furthermore, the presence of detrital or early diagenetic kaolinite coatings in fluvio-eolian samples can be shown. However, this indication is just based on 60 samples collected at approximately 20 cm below the surface where undisturbed sedimentary textures were observed.

Although samples were collected across multiple positions within one dune, across multiple dunes per area and also from other depositional environments, the author does note the necessity for an even broader database to acquire more detailed data in the presented and additional areas. Should this topic be followed, the author advises the sampling to be performed in form of continuous sediment cores to allow the delineation of continuous trends throughout the deposits. Groundwater, if available, should be sampled to determine the stability of clay minerals in these solutions. This could indicate more detailed geochemical conditions during emplacement and formation of clay coatings in active depositional systems.

The further incorporation of depositional environments outside of the arid to semi-arid fluvio-eolian environment should be considered. The systematic study of arid and humid fluvial samples, as well as marine samples could contain dependencies which can be delineated based on the depositional environment. More stable pore fluid-saturations and -chemistries as opposed to the arid to semi-arid fluvio-eolian of the Rotliegend might result in a more systematic occurrence of grain coatings.

The viability of using quartz and illite kinetic models to validate burial models calibrated by organic matter maturation data could be tested on more complex settings, where organic matter maturation data is rare, or where a complex organic matter maturation pattern is known. Additionally, the algorithms can be tested in a hydrothermally influenced area to highlight the added value of kinetic illite and quartz cementation modeling.

Reservoir quality analysis and modeling workflow could be eased by automatically detecting grain coat coverages based on light microscopic analyses, which would reduce the petrographers bias and increase accuracy.

7 References

Aagaard, P., J. S. Jahren, A. O. Harstad, O. Nilsen, and M. Ramm, 2000, Formation of grain-coating chlorite in sandstones. Laboratory synthesized vs. natural occurrences: Clay Minerals, v. 35, p. 261-269, doi:10.1180/000985500546639.

Ahmadov, R., A. Aydin, M. Karimi-Fard, and L. J. Durlofsky, 2007, Permeability upscaling of fault zones in the Aztec Sandstone, Valley of Fire State Park, Nevada, with a focus on slip surfaces and slip bands: Hydrogeology Journal, v. 15, p. 1239-1250, doi:10.1007/s10040-007-0180-2.

Ajdukiewicz, J. M., and R. H. Lander, 2010, Sandstone reservoir quality prediction: The state of the art: AAPG Bulletin, v. 94, p. 1083-1091, doi:10.1306/intro060110.

Ajdukiewicz, J. M., P. H. Nicholson, and W. L. Esch, 2010, Prediction of deep reservoir quality using early diagenetic process models in the Jurassic Norphlet Formation, Gulf of Mexico: AAPG Bulletin, v. 94, p. 1189-1227, doi:10.1306/04211009152.

Ajdukiewicz, J. M., and R. E. Larese, 2012, How clay grain coats inhibit quartz cement and preserve porosity in deeply buried sandstones: Observations and experiments: AAPG Bulletin, v. 96, p. 2091-2119, doi:10.1306/02211211075.

Ajdukiewicz, J. M., K. Kronmüller, C. Dulce, and G. Yip, 2014, Integrating early and late diagenesis models for spatial prediction of reservoir quality in deeply-buried eolian sandstones (abs.): AAPG International Conference & Exhibition.

Allen, P. A., and J. R. Allen, 2005, Basin Analysis - Principles and Applications: Malden, Blackwell Publishing Ltd., 560 p.

Amthor, J. E., and J. Okkerman, 1998, Influence of early diagenesis on reservoir quality of Rotliegende sandstones, Northern Netherlands: AAPG Bulletin, v. 82, p. 2246-2265, doi:10.1306/00AA7F04-1730-11D7-8645000102C1865D.

Anderson, R. S., 1988, The pattern of grainfall deposition in the lee of aeolian dunes: Sedimentology, v. 35, p. 175-188, doi:10.1111/j.1365-3091.1988.tb00943.x.

Anjos, S. M. C., L. F. De Ros, R. Schiffer de Souza, C. M. A. Silva, and C. L. Sombra, 2000, Depositional and diagenetic controls on the reservoir quality of Lower Cretaceous Pendência sandstones, Potiguar rift basin, Brazil: AAPG Bulletin, v. 84, p. 1719-1742, doi:10.1306/8626C375-173B-11D7-8645000102C1865D.

Anjos, S. M. C., L. F. De Ros, and C. M. A. Silva, 2003, Chlorite authigenesis and porosity preservation in the Upper Cretaceous marine sandstones of the Santos Basin, offshore eastern Brazil, *in* R. H. Worden, and S. Morad, eds., Clay Mineral Cements in Sandstones: Oxford, Blackwell Science Ltd., doi:10.1002/9781444304336.ch13.

Antonellini, M., and A. Aydin, 1994, Effect of faulting on fluid flow in porous sandstones: Petrophysical properties: AAPG Bulletin, v. 78, p. 355-377.

Antonellini, M., and A. Aydin, 1995, Effect of faulting on fluid flow in porous sandstones: Geometry and spatial distribution: AAPG Bulletin, v. 79, p. 642-671.

Antrett, P., 2011, Characterization of an Upper Permian Tight Gas Reservoir - A multidisciplinary, multiscale analysis from the Rotliegend, Northern Germany: PhD thesis, RWTH Aachen University, Aachen, 125 p.

Armitage, P. J., D. R. Faulkner, R. H. Worden, A. C. Aplin, A. R. Butcher, and J. Iliffe, 2011, Experimental measurement of, and controls on, permeability and permeability anisotropy of caprocks from the CO2 storage project at the Krechba Field, Algeria: Journal of Geophysical Research, v. 116, p. B12208, doi:10.1029/2011jbC08385.

Armitage, P. J., R. H. Worden, D. R. Faulkner, A. C. Aplin, A. R. Butcher, and A. A. Espie, 2013, Mercia Mudstone Formation caprock to carbon capture and storage sites: petrology and petrophysical characteristics: Journal of the Geological Society, v. 170, p. 119-132, doi:10.1144/jgs2012-049.

Arthurton, R. S., and A. J. Wadge, 1981, Geology of the country around Penrith, Mem. Geol. Surv. G.B., p. 177.

Awdal, A., D. Healy, and G. I. Alsop, 2014, Geometrical analysis of deformation band lozenges and their scaling relationships to fault lenses: Journal of Structural Geology, v. 66, p. 11-23, doi:10.1016/j.jsg.2014.05.006.

Awwiller, D. N., and L. L. Summa, 1997, Quartz cement volume constraints on burial history analysis: An example from the Eocene of western Venezuela (abs.): AAPG Annual Meeting Program, v. 6, p. A6.

Bahlis, A. B., and L. F. De Ros, 2013, Origin and impact of authigenic chlorite in the Upper Cretaceous sandstone reservoirs of the Santos Basin, eastern Brazil: Petroleum Geoscience, v. 19, p. 185-199, doi:10.1144/petgeo2011-007.

Baldschuhn, R., F. Binot, S. Fleig, and F. Kockel, 2001, Geotektonischer Atlas von Nordwest-Deutschland und dem deutschen Nordsee-Sektor: Geologisches Jahrbuch, v. 153: Hannover, E. Schweizerbart'sche Verlagsbuchhandlung, Stuttgart, 88 p.

Ballas, G., R. Soliva, J.-P. Sizun, A. Benedicto, T. Cavailhes, and S. Raynaud, 2012, The importance of the degree of cataclasis in shear bands for fluid flow in porous sandstone, Provence, France: AAPG Bulletin, v. 96, p. 2167-2186, doi:10.1306/04051211097.

Baud, P., P. Meredith, and E. Townend, 2012, Permeability evolution during triaxial compaction of an anisotropic porous sandstone: Journal of Geophysical Research, v. 117, p. B05203, doi:10.1029/2012jb009176.

Benson, P. M., P. G. Meredith, E. S. Platzman, and R. E. White, 2005, Pore fabric shape anisotropy in porous sandstones and its relation to elastic wave velocity and permeability anisotropy under hydrostatic pressure: International Journal of Rock Mechanics and Mining Sciences, v. 42, p. 890-899, doi:10.1016/j.ijrmms.2005.05.003.

Bésuelle, P., and J. W. Rudnicki, 2004, Localization: Shear bands and compaction bands, *in* Y. Gueguen, and M. Boutéca, eds., Mechanics of Fluid-Saturated Rocks: Amsterdam, Elsevier Academic Press, p. 219-321.

BGS, 2016, The BGS Lexicon of Named Rock Units, http://www.bgs.ac.uk/lexicon/lexicon.cfm?pub=PS, 14.04.2016

Billault, V., D. Beaufort, A. Barnonnet, and J.-C. Lacharpagne, 2003, A nanopetrographic and textural study of grain-coating chlorites in sandstone reservoirs: Clay Minerals, v. 38, p. 315-328, doi:10.1180/0009855033830098.

Bjørlykke, K., 2014, Relationships between depositional environments, burial history and rock properties. Some principal aspects of diagenetic process in sedimentary basins: Sedimentary Geology, v. 301, p. 1-14, doi:10.1016/j.sedgeo.2013.12.002.

Bloch, S., R. H. Lander, and L. M. Bonnell, 2002, Anomalously high porosity and permeability in deeply buried sandstone reservoirs: Origin and predictability: AAPG Bulletin, v. 86, p. 301-328, doi:10.1306/61EEDABC-173E-11D7-8645000102C1865D.

Boles, J. R., and S. G. Franks, 1979, Clay diagenesis in Wilcox sandstones of Southwest Texas; implications of smectite diagenesis on sandstone cementation: Journal of Sedimentary Petrology, v. 49, p. 55-70, doi:10.1306/212F76BC-2B24-11D7-8648000102C1865D.

Bombien, H., B. Hoffers, S. Breuckmann, M. Helms, K. Lademann, M. Lange, A. Oelrich, R. Reimann, J. Rienäcker, C. Schmidt, M.-F. Slaby, and J. Ziesch, 2012, Der Geotektonische Atlas von Niedersachsen und dem deutschen Nordseesektor als geologisches 3D-Modell: GMIT, p. 6-13.

Bruns, B., R. di Primio, U. Berner, and R. Littke, 2013, Petroleum system evolution in the inverted Lower Saxony Basin, northwest Germany: a 3D basin modeling study: Geofluids, v. 13, p. 246-271, doi:10.1111/gfl.12016.

Bukar, M., 2013, Does oil emplacement stop diagenesis and quartz cementation in deeply buried sandstone reservoirs: PhD thesis, University of Liverpool, Liverpool, 270 p.

Burnham, A. K., and J. J. Sweeney, 1989, A chemical kinetic model of vitrinite maturation and reflectance: Geochimica et Cosmochimica Acta, v. 53, p. 2649-2687, doi:10.1016/0016-7037(89)90136-1.

Busch, B., R. Winkler, K. Osivandi, G. Nover, A. Amann-Hildenbrand, and C. Hilgers, 2015, Evolution of small-scale flow barriers in German Rotliegend siliciclastics, *in* P. J. Armitage, A. R. Butcher, J. M. Churchill, A. E. Csoma, C. Hollis, R. H. Lander, J. E. Omma, and R. H. Worden, eds., Reservoir Quality of Clastic and Carbonate Rocks: Analysis, Modelling and Prediction, Geological Society, London, Special Publication, doi:10.1144/sp435.3.

Chemenda, A. I., C. Wibberley, and E. Saillet, 2012, Evolution of compactive shear deformation bands: Numerical models and geological data: Tectonophysics, v. 526-529, p. 56-66, doi:10.1016/j.tecto.2011.10.003.

Cheung, C. S. N., P. Baud, and T.-f. Wong, 2012, Effect of grain size distribution on the development of compaction localization in porous sandstone: Geophysical Research Letters, v. 39, p. L21302, doi:10.1029/2012GL053739.

Darbha, G. K., C. Fischer, J. Luetzenkirchen, and T. Schafer, 2012, Site-specific retention of colloids at rough rock surfaces: Environmental science & technology, v. 46, p. 9378-87, doi:10.1021/es301969m.

Davatzes, N. C., and A. Aydin, 2003, Overprinting faulting mechanisms in high porosity sandstones of SE Utah: Journal of Structural Geology, v. 25, p. 1795-1813, doi:10.1016/s0191-8141(03)00043-9.

David, C., T.-f. Wong, W. Zhu, and J. Zhang, 1994, Laboratory measurement of compaction-induced permeability change in porous rocks: Implications for the generation and maintenance of pore pressure excess in the crust: Pure and Applied Geophysics, v. 143, p. 425-456, doi:10.1007/BF00874337.

David, C., B. Menendez, W. Zhu, and T.-f. Wong, 2001, Mechanical compaction, microstructures and permeability evolution in sandstones: Phys. Chem. Earth (A), v. 26, p. 45-51, doi:10.1016/S1464-1895(01)00021-7.

De Ros, L. F., S. M. C. Anjos, and S. Morad, 1994, Authigenesis of amphibole and its relationship to the diagenetic evolution of Lower Cretaceous sandstones of the Potiguar rift basin, northeastern Brazil: Sedimentary Geology, v. 88, p. 253-266, doi:10.1016/0037-0738(94)90065-5.

Deng, S., L. Zuo, A. Aydin, J. Dvorkin, and T. Mukerji, 2015, Permeability characterization of natural compaction bands using core flooding experiments and three-dimensional image-based analysis: Comparing and contrasting the results from two different methods: AAPG Bulletin, v. 99, p. 27-49, doi:10.1306/07071413211.

Deutsche Stratigraphische Kommission, 2002, Stratigraphische Tabelle von Deutschland 2002, v. 1, Deutsche Stratigraphische Kommission (DSK); Potsdam (GeoForschungsZentrum); Frankfurt a. M. (Forsch.-Inst. Senckenberg).

Dowey, P. J., D. M. Hodgson, and R. H. Worden, 2012, Pre-requisites, processes, and prediction of chlorite grain coatings in petroleum reservoirs: A review of subsurface examples: Marine and Petroleum Geology, v. 32, p. 63-75, doi:10.1016/j.marpetgeo.2011.11.007.

Ehrenberg, S. N., and P. H. Nadeau, 1989, Formation of diagenetic illite in sandstones of the Garn Formation, Haltenbanken area, Mid-Norwegian continental shelf: Clay Minerals, v. 24, p. 233-253, doi:10.1306/BDFF8E5C-1718-11D7-8645000102C1865D.

Ehrenberg, S. N., 1993, Preservation of anomalously high porosity in deeply buried sandstones by grain-coating chlorite: Examples from the Norwegian continental shelf: AAPG Bulletin, v. 77, p. 1260-1286, doi:10.1306/BDFF8E5C-1718-11D7-8645000102C1865D.

Eichhubl, P., W. L. Taylor, D. D. Pollard, and A. Aydin, 2004, Paleo-fluid flow and deformation in the Aztec Sandstone at the Valley of Fire, Nevada - Evidence for the coupling of hydrogeologic, diagenetic, and tectonic processes: Geological Society of America Bulletin, v. 116, p. 1120-1131, doi:10.1130/b25446.1.

Eichhubl, P., and E. Flodin, 2005, Brittle deformation, fluid flow, and diagenesis in sandstones at Valley of Fire State Park, Nevada, *in* J. Pederson, and C. M. Dehler, eds., Interior Western United States: Geological Society of America field guide 6, Geological Society of America, p. 151-167, doi:10.1130/2005.fl.

El-Ghali, M. A. K., H. Mansurbeg, S. Morad, I. Al-Aasm, and G. Ajdanlisky, 2006, Distribution of diagenetic alterations in fluvial and paralic deposits within sequence stratigraphic framework: Evidence from the Petrohan Terrigenous Group and the Svidol Formation, Lower Triassic, NW Bulgaria: Sedimentary Geology, v. 190, p. 299-321, doi:10.1016/j.sedgeo.2006.05.021.

El-Gomati, M. M., C. G. H. Walker, and X. Zha, 2011, Towards quantitative scanning electron microscopy: Applications to nano-scale analysis: Nuclear Instruments and Methods in Physics Research Section A: Accelerators, Spectrometers, Detectors and Associated Equipment, v. 645, p. 68-73, doi:10.1016/j.nima.2010.12.133.

Elias, A. R. D., L. F. De Ros, A. M. P. Mizusaki, and S. M. C. Anjos, 2004, Diagenetic patterns in eolian/coastal sabkha reservoirs of the Solimões Basin, Northern Brazil: Sedimentary Geology, v. 169, p. 191-217, doi:10.1016/j.sedgeo.2004.06.001.

Esch, W. L., J. M. Ajdukiewicz, and A. C. Reynolds, 2008, Early grain-coat formation in Chaco dune field, New Mexico: insight into formation mechanisms, distribution, and implications for predictive modeling to assist in deep play identification (abs): AAPG Annual Convention.

Exner, U., J. Kaiser, and S. Gier, 2013, Deformation bands evolving from dilation to cementation bands in a hydrocarbon reservoir (Vienna Basin, Austria): Marine and Petroleum Geology, v. 43, p. 504-515, doi:10.1016/j.marpetgeo.2012.10.001.

Faulkner, D. R., and E. H. Rutter, 2000, Comparisons of water and argon permeability in natural clay-bearing fault gouge under high pressure at 20°C: Journal of Geophysical Research v. 105, p. 16415-16426, doi:10.1029/2000JB900134.

Fischer, C., R. Gaupp, M. Dimke, and O. Sill, 2007, A 3D High Resolution model of bounding surfaces in aeolian-fluvial deposits: An outcrop analogue study from the Permian Rotliegend, northern Germany: Journal of Petroleum Geology, v. 30, p. 257-274, doi:10.1111/j.1747-5457.2007.00257.x.

Fischer, C., I. Dunkl, H. von Eynatten, J. R. Wijbrans, and R. Gaupp, 2012, Products and timing of diagenetic processes in Upper Rotliegend sandstones from Bebertal (North German Basin, Parchim Formation, Flechtingen High, Germany): Geological Magazine, v. 149, p. 827-840, doi:10.1017/s0016756811001087.

Fischer, C., P. Aurin, G. K. Darbha, and G. Arp, 2013, Experimental approaches to the formation of early-diagenetic grain coats on quartz surfaces: Zeitschrift der Deutschen Gesellschaft für Geowissenschaften, v. 164, p. 225-236, doi:10.1127/1860-1804/2013/0030.

Fisher, Q. J., R. J. Knipe, and R. H. Worden, 2000, Microstructures of deformed and non-deformed sandstones from the North Sea: implications for the origins of quartz cement in sandstones, *in* R. H. Worden, and S. Morad, eds., Quartz Cementation in Sandstones, v. 29: Oxford, Blackwell Science Ltd., p. 129-146, doi:10.1002/9781444304237.ch10.

Fisher, Q. J., and R. Knipe, 2001, Permeability of faults within siliciclastic petroleum reservoirs of the North Sea and Norwegian continental shelf: Marine and Petroleum Geology, v. 18, p. 1063-1081.

Folk, R. L., and W. Ward, 1957, Brazos River bar: a study in the significance of grain-size parameters: Journal of Sedimentary Petrology, v. 27, p. 3-26, doi:10.1306/74D70646-2B21-11D7-8648000102C1865D.

Folk, R. L., 1980, Petrology of Sedimentary Rocks: Austin, Texas, U.S.A., Hemphill Publishing Company.

Fossen, H., and A. Bale, 2007, Deformation bands and their influence on fluid flow: AAPG Bulletin, v. 91, p. 1685-1700, doi:10.1306/07300706146.

Fossen, H., R. A. Schultz, Z. K. Shipton, and K. Mair, 2007, Deformation bands in sandstone: A review: Journal of the Geological Society, London, v. 164, p. 1-15, doi:10.1144/0016-76492006-036.

Fossen, H., 2010, Deformation bands formed during soft-sediment deformation: Observations from SE Utah: Marine and Petroleum Geology, v. 27, p. 215-222, doi:10.1016/j.marpetgeo.2009.06.005.

Fossen, H., R. A. Schultz, and A. Torabi, 2011, Conditions and implications for compaction band formation in the Navajo Sandstone, Utah: Journal of Structural Geology, v. 33, p. 1477-1490, doi:10.1016/j.jsg.2011.08.001.

Fossen, H., and A. Rotevatn, 2012, Characterization of deformation bands associated with normal and reverse stress states in the Navajo Sandstone, Utah: Discussion: AAPG Bulletin, v. 96, p. 869-876, doi:10.1306/09221110173.

Fowler, C. M., D. Stead, B. I. Pandit, B. W. Janser, E. G. Nisbet, and G. Nover, 2005, A database of physical properties of rocks from the Trans-Hudson Orogen, Canada: Canadian Journal of Earth Sciences, v. 42, p. 555-572, doi:10.1139/e05-006.

Fowles, J., and S. D. Burley, 1994, Textural and permeability characteristics of faulted, high porosity sandstones: Marine and Petroleum Geology, v. 11, p. 608-623, doi:10.1016/0264-8172(94)90071-X.

French, M. W., R. H. Worden, E. Mariani, R. E. Larese, R. R. Mueller, and C. E. Kliewer, 2012, Microcrystalline quartz generations and the preservation of porosity in sandstones: Evidence from the Upper Cretaceous of the Subhercynian Basin, Germany: Journal of Sedimentary Research, v. 82, p. 422-434, doi:10.2110/jsr.2012.39.

Freund, D., and G. Nover, 1995, Hydrostatic pressure tests for the permeability-formation factor relation on crystalline rocks from KTB drilling project: Surveys in Geophysics, v. 16, p. 47-62, doi: 10.1007/BF00682712.

Fryberger, S. G., R. Knight, C. Hern, A. Moscariello, and S. Kabel, 2011, Rotliegend facies, sedimentary provinces, and stratigraphy, Southern Permian Basin UK and the Netherlands: A review with new observations, *in* J. Grötsch, and R. Gaupp, eds., The Permian Rotliegend of the Netherlands, v. 98: Tulsa, Oklahoma, U.S.A., Society for Sedimentary Geology, p. 51-88, doi:10.2110/pec.11.98.0051.

Gaitzsch, B., M. Geissler, M. Göthel, J. Kopp, and B. Legler, 2004, Karbon und Rotliegend der Flechtingen-Roßlau-Scholle, *in* C. H. Friedel, and H. G. Röhling, eds., GeoLeipzig 2004 - Geowissenschaften sichern Zukunft. 29.September - 1.Oktober 2004 in Leipzig, Exkursionsführer. Schriftenreihe Deutsche Geologische Gesellschaft, v. 35, p. 150-164.

GAPS, 1982, Sedimentological and petrological study of Rotliegend (cores 1-5) from well K/10-B04, London, UK, GAPS Geological Services.

GAPS, 1983, Sedimentological and petrological study of cores 1, 2, 3, and 4 (3264-3317.93m) from wll K/10-C2 London, UK, GAPS Geological Services.

GAPS, 1984a, Facies analysis and reservoir evaluation of the Rotliegend in Dutch blocks K/10 and K/13, London, UK, GAPS Geological Services.

GAPS, 1984b, Sedimentological and Petrological study of cores 1, 2, 3, and 4 (2479-2550.5m) from well K/13-4, London, UK, GAPS Geological Services.

GAPS, 1994, Sedimentology, petrology, and reservoir properties of the Upper Rotliegend Deposits in cores 1 to 7 from well K10-V2, Warmond, The Netherlands, GAPS Nederland B.V.

Gast, R. E., M. Dusar, C. Breitkreuz, R. Gaupp, J. W. Schneider, L. Stemmerik, M. C. Geluk, M. Geißler, H. Kiersnowski, K. W. Glennie, S. Kabel, and N. S. Jones, 2010, Rotliegend, *in* J. C. Doornenbal, and A. G. Stevenson, eds., Petroleum Geological Atlas of the Southern Permian Basin Area: Houten, EAGE Publications BV, p. 101-123.

Gaupp, R., A. Matter, J. D. Platt, K. Ramseyer, and J. Walzebuck, 1993, Diagenesis and fluid evolution of deeply buried Permian (Rotliegend) gas reservoirs, northwest Germany: AAPG Bulletin, v. 77, p. 1111-1128, doi:10.1306/BDFF8E0C-1718-11D7-8645000102C1865D.

Gier, S., R. H. Worden, W. D. Johns, and H. Kurzweil, 2008, Diagenesis and reservoir quality of Miocene sandstones in the Vienna Basin, Austria: Marine and Petroleum Geology, v. 25, p. 681-695, doi:10.1016/j.marpetgeo.2008.06.001.

Glennie, K. W., 1983, Lower Permian Rotliegend desert sedimentation in the North Sea area, *in* M. E. Brookfield, and T. S. Ahlbrandt, eds., Eolian Sediments and Processes, v. 32: Amsterdam, Elsevier Science Publishers B.V., p. 521-541.

Griffiths, J., R. H. Worden, D. R. Faulkner, and A. Edwards, 2016, Deformation band character as a function of grain size, mineralogy and porosity in Triassic Sherwood Sandstone, *in* P. J. Armitage, A. Butcher, A. Csoma, C. Hollis, R. H. Lander, J. Omma, and R. H. Worden, eds., Reservoir Quality of Clastic and Carbonate Rocks: Analysis, Modelling, and Prediction: Special Publications, v. 435: London, Geological Society, doi:10.1144/SP435.11.

Grigsby, J. D., 2001, Origin and growth mechanism of authigenic chlorite in sandstones of the lower Vicksburg Formation, South Texas: Journal of Sedimentary Research, v. 71, p. 27-36, doi:10.1306/060100710027.

Guo, J., K. McCaffrey, R. Jones, and R. Holdsworth, 2009, The spatial heterogeneity of structures in high porosity sandstones: Variations and granularity effects in orientation data: Journal of Structural Geology, v. 31, p. 628-636, doi:10.1016/j.jsg.2008.12.002.

Haile, B. G., H. Hellevang, P. Aagaard, and J. Jahren, 2015, Experimental nucleation and growth of smectite and chlorite coatings on clean feldspar and quartz grain surfaces: Marine and Petroleum Geology, v. 68, p. 664-674, doi:10.1016/j.marpetgeo.2015.02.006.

Hammer, E., M. B. E. Mørk, and A. Næss, 2010, Facies controls on the distribution of diagenesis and compaction in fluvial-deltaic deposits: Marine and Petroleum Geology, v. 27, p. 1737-1751, doi:10.1016/j.marpetgeo.2009.11.002.

Harwood, J., A. C. Aplin, C. I. Fialips, J. Iliffe, R. Kozdon, T. Ushikubo, and J. W. Valley, 2013, Quartz cementation history of sandstone revealed by high-resolution SIMS oxygen isotope analysis: Journal of Sedimentary Research, v. 83, p. 522-530, doi:10.2110/jsr.2013.29.

Heald, M. T., and R. E. Larese, 1974, Influence of coatings on quartz cementation: Journal of Sedimentary Petrology, v. 44, p. 1269-1274, doi:10.1306/212F6C94-2B24-11D7-8648000102C1865D.

Henares, S., L. Caracciolo, G. Cultrone, J. Fernández, and C. Viseras, 2014, The role of diagenesis and depositional facies on pore system evolution in a Triassic outcrop analogue (SE Spain): Marine and Petroleum Geology, v. 51, p. 136-151, doi:10.1016/j.marpetgeo.2013.12.004.

Hildenbrand, A., S. Schlömer, and B. M. Krooss, 2002, Gas breakthrough experiments on fine-grained sedimentary rocks: Geofluids, v. 2, p. 3-23, doi:10.1046/j.1468-8123.2002.00031.x.

Hillier, S., 1994, Pore-lining chlorites in siliciclastic reservoir sandstones: Electron microprobe, SEM and XRD data, and implications for their origin: Clay Minerals, v. 29, p. 665-679.

Holt, R. M., 1994, Effects of coring on petrophysical measurements: SCA International Symposium of the Society of Core Analysts, p. 77-86.

Houseknecht, D. W., 1987, Assessing the relative importance of compaction processes and cementation to reduction of porosity in sandstones: AAPG Bulletin, v. 71, p. 633-642.

Hughes, R. A., 2003, Permian and Triassic rocks of the Appleby district (part of Sheet 30, England and Wales), *in* B. G. Survey, ed., Research Report, Keyworth, Nottingham, British Geological Survey, p. 21.

Johansen, T. E. S., H. Fossen, and R. Kluge, 2005, The impact of syn-faulting porosity reduction on damage zone architecture in porous sandstone: an outcrop example from the Moab Fault, Utah: Journal of Structural Geology, v. 27, p. 1469-1485, doi:10.1016/j.jsg.2005.01.014.

Johansen, T. E. S., and H. Fossen, 2008, Internal geometry of fault damage zones in interbedded siliciclastic sediments, *in* C. A. J. Wibberley, W. Kurz, J. Imber, R. E. Holdsworth, and C. Collettini, eds., The Internal Structure of Fault Zones: Implications for Mechanical and Fluid-Flow Properties, Geological Society, London, Special Publications, p. 35-56, doi:10.1144/sp299.3.

Jones, F. O., and W. W. Owens, 1980, A laboratory study of low-permeability gas sands: Journal of Petroleum Technology, v. 32, p. 1631-1640, doi:10.2118/7551-PA.

Kayser, A., 2006, Herkunft, Auftreten und Visualisierung von Permeabilitätsbarrieren in einer Gaslagerstätte in Sandsteinen des Rotliegenden (Südliches Permbecken, Deutschland): PhD thesis, Philipps-Universität Marburg, Marburg, 206 p.

Ketzer, J. M., S. Morad, E. R., and I. S. Al-Aasm, 2002, Distribution of diagenetic alterations in fluvial, deltaic, and shallow marine sandstones, within a sequence stratigraphic framework: Journal of Sedimentary Research, v. 72, p. 760-774, doi:10.1306/042202720760.

Khalifa, M. A., and S. Morad, 2015, Impact of depositional facies on the distribution of diagenetic alterations in the Devonian shoreface sandstone reservoirs, Southern Ghadamis Basin, Libya: Sedimentary Geology, v. 329, p. 62-80, doi:10.1016/j.sedgeo.2015.09.003.

Kley, J., H.-J. Franzke, F. Jähne, C. Krawczyk, T. Lohr, K. Reicherter, M. Scheck-Wenderoth, J. Sippel, D. Tanner, and H. van Gent, 2008, Strain and stress, *in* R. Littke, U. Bayer, D. Gajewski, and S. Nelskamp, eds., Dynamics of complex intracontinental basins – The Central European Basin System: Berlin Heidelberg, Springer, p. 97–124.

Kolyukhin, D., S. Schueller, M. S. Espedal, and H. Fossen, 2009, Deformation band populations in fault damage zone - impact on fluid flow: Computational Geosciences, v. 14, p. 231-248, doi:10.1007/s10596-009-9148-8.

Lan, C., M. Yang, and Y. Zhang, 2016, Impact of sequence stratigraphy, depositional facies and diagenesis on reservoir quality: A case study on the Pennsylvanian Taiyuan sandstones, northeastern Ordos Basin, China: Marine and Petroleum Geology, v. 69, p. 216-230, doi:10.1016/j.marpetgeo.2015.09.009.

Lander, R. H., V. Felt, L. M. Bonnell, and O. Walderhaug, 1997a, Utility of sandstone diagenetic modeling for basin history assessment (abs.): AAPG Annual Meeting Program, v. 6, p. A66.

Lander, R. H., O. Walderhaug, and L. M. Bonnell, 1997b, Application of sandstone diagenetic modeling to reservoir quality prediction and basin history assessment, Memorias del I Congreso Latinoamericano de Sedimentología, Caracas, Venezuela, Sociedad Venezolana de Geólogos, p. 373-386.

Lander, R. H., and O. Walderhaug, 1999, Predicting porosity through simulating sandstone compaction and quartz cementation: AAPG Bulletin, v. 83, p. 433-449, doi:10.1306/00AA9BC4-1730-11D7-8645000102C1865D.

Lander, R. H., R. E. Larese, and L. M. Bonnell, 2008, Toward more accurate quartz cement models: The importance of euhedral versus noneuhedral growth rates: AAPG Bulletin, v. 92, p. 1537-1563, doi:10.1306/07160808037.

Lander, R. H., and L. M. Bonnell, 2010, A model for fibrous illite nucleation and growth in sandstones: AAPG Bulletin, v. 94, p. 1161-1187, doi:10.1306/04211009121.

Lander, R. H., and S. E. Laubach, 2014, Insights into rates of fracture growth and sealing from a model for quartz cementation in fractured sandstones: Geological Society of America Bulletin, v. 127, p. 516-538, doi:10.1130/b31092.1.

Lanson, B., D. Beaufort, G. Berger, A. Bauer, A. Cassagnabère, and A. Meunier, 2002, Authigenic kaolin and illitic minerals during burial diagenesis of sandstones: a review: Clay Minerals, v. 37, p. 1-22, doi:10.1180/0009855023710014.

Le Bayon, R., G. P. Brey, W. G. Ernst, and R. F. Mählmann, 2011, Experimental kinetic study of organic matter maturation: Time and pressure effects on vitrinite reflectance at 400°C: Organic Geochemistry, v. 42, p. 340-355, doi:10.1016/j.orggeochem.2011.01.011.

Le Bayon, R., S. Buhre, B. C. Schmidt, and R. F. Mählmann, 2012, Experimental organic matter maturation at 2kbar: Heat-up effect to low temperatures on vitrinite reflectance: International Journal of Coal Geology, v. 92, p. 45-53, doi:10.1016/j.coal.2011.12.002.

Legler, B., and C. Marchel, 2008, Deformation bands in Permian reservoir sandstones (Rotliegend, Northwest Germany): Freiberger Forschungshefte, v. C528, p. 125-138.

Liewig, N., and N. Clauer, 2000, K-Ar dating of varied microtextural illite in Permian gas reservoirs, northern Germany: Clay Minerals, v. 35, p. 271-281, doi:10.1180/000985500546648.

Littke, R., M. Scheck-Wenderoth, M. R. Brix, and S. Nelskamp, 2008, Subsidence, inversion and evolution of the thermal field, *in* R. Littke, U. Bayer, D. Gajewski, and S.

Nelskamp, eds., Dynamics of complex intracontinental basins – The Central European Basin System: Berlin, Springer, p. 125-153.

Lohr, T., C. M. Krawczyk, D. C. Tanner, R. Samiee, H. Endres, O. Oncken, H. Trappe, and P. A. Kukla, 2007, Strain partitioning due to salt: insights from interpretation of a 3D seismic data set in the NW German Basin: Basin Research, v. 19, p. 579-597, doi:10.1111/j.1365-2117.2007.00338.x.

Louis, L., C. David, V. Metz, P. Robion, B. Menéndez, and C. Kissel, 2005, Microstructural control on the anisotropy of elastic and transport properties in undeformed sandstones: International Journal of Rock Mechanics and Mining Sciences, v. 42, p. 911-923, doi:10.1016/j.ijrmms.2005.05.004.

Macchi, L. C., 1981, Sedimentology of the Penrith Sandstone and brockrams (Permo-Triassic) of Cumbria, north-west England: Volume 1: PhD thesis, University of Hull, Hull, 411 p.

MacDonald, R., L. Wilson, R. S. Thorpe, and A. Martin, 1988, Emplacement of the Cleveland Dyke: Evidence from geochemistry, mineralogy and physical modelling: Journal of Petrology, v. 29, p. 559-583, doi:10.1093/petrology/29.3.559.

Mader, D., and M. J. Yardley, 1985, Migration, modification and merging in aeolian systems and the significance of the depositional mechanisms in Permian and Triassic dune sands of Europe and North America: Sedimentary Geology, v. 43, p. 85-218, doi:10.1016/0037-0738(85)90056-9.

Mahaney, W. C., D. I. Netoff, J. Dohm, R. G. V. Hancock, and D. Krinsley, 2010, Grain coatings: Diagenesis of Jurassic sandstones in south-central Utah and implications for targeting fossil microbes on Mars: Sedimentary Geology, v. 230, p. 1-9, doi:10.1016/j.sedgeo.2010.06.015.

Main, I. G., O. Kwon, B. T. Ngwenya, and S. C. Elphick, 2000, Fault sealing during deformation-band growth in porous sandstone: Geology, v. 28, p. 1131-1134, doi:10.1130/0091-7613(2000)28<1131:FSDDGI>2.0.CO;2.

Makowitz, A., R. H. Lander, and K. L. Milliken, 2006, Diagenetic modeling to assess the relative timing of quartz cementation and brittle grain processes during compaction: AAPG Bulletin, v. 90, p. 873-885, doi:10.1306/12190505044.

Marchand, A. M. E., G. Apps, W. Li, and J. R. Rotzien, 2015, Depositional processes and impact on reservoir quality in deepwater Paleogene reservoirs, US Gulf of Mexico: AAPG Bulletin, v. 99, p. 1635-1648, doi:10.1306/04091514189.

Matthews, J. M., J. Samuelson, C. Lowrey, and H. M. Helset, 2002, Quartz cement abundance as a thermal history indicator: - A case study from the Gjallar Ridge, Vøring Basin (abs.), AAPG Hedberg Conference, Stavanger, Norway, AAPG.

McKenzie, D., 1978, Some remarks on the development of sedimentary basins: Earth and Planetary Science Letters, v. 40, p. 25-32, doi:10.1016/0012-821X(78)90071-7.

Menning, M., 1995, A numerical time scale for the Permian and Triassic periods: An integrated time analysis, *in* P. A. Scholle, T. M. Peryt, and D. S. Ulmer-Scholle, eds., The Permian of Northern Pangea, Volume 1: Paleogeography, paleoclimates, stratigraphy, v. 1: Berlin, Springer, p. 77-97.

Morad, S., J. M. Ketzer, and L. F. De Ros, 2000, Spatial and temporal distribution of diagenetic alterations in siliciclastic rocks: implications for mass transfer in sedimentary basins: Sedimentology, v. 47, p. 95-120, doi:10.1046/j.1365-3091.2000.00007.x.

Morad, S., K. Al-Ramadan, J. M. Ketzer, and L. F. De Ros, 2010, The impact of diagenesis on the heterogeneity of sandstone reservoirs: A review of the role of depositional facies and sequence stratigraphy: AAPG Bulletin, v. 94, p. 1267-1309, doi:10.1306/04211009178.

Moraes, M. A. S., and L. F. De Ros, 1992, Depositional, infiltrated and authigenic clays in fluvial sandstones of the Jurassic Sergi Formation, Reconcavo Basin, northeastern Brazil, *in* D. W. Houseknecht, and E. D. Pittman, eds., Origin, Diagenesis, and Petrophysics of Clay Minerals in Sandstones: Tulsa, Oklahoma, U.S.A., SEPM, doi:10.2110/pec.92.47.0197.

Mountney, N. P., and A. Jagger, 2004, Stratigraphic evolution of an aeolian erg margin system: the Permian Cedar Mesa Sandstone, SE Utah, USA: Sedimentology, v. 51, p. 713-743, doi:10.1111/j.1365-3091.2004.00646.x.

Nicol, A., C. Childs, J. J. Walsh, and K. W. Schafer, 2013, A geometric model for the formation of deformation band clusters: Journal of Structural Geology, v. 55, p. 21-33, doi:10.1016/j.jsg.2013.07.004.

Nixon, C. W., D. J. Sanderson, S. J. Dee, J. M. Bull, R. J. Humphreys, and M. H. Swanson, 2014, Fault interactions and reactivation within a normal-fault network at Milne Point, Alaska: AAPG Bulletin, v. 98, p. 2081-2107, doi:10.1306/04301413177.

Ogilvie, S. R., and P. W. J. Glover, 2001, The petrophysical properties of deformation bands in relation to their microstructure: Earth and Planetary Science Letters, v. 193, p. 123-142, doi:10.1016/S0012-821X(01)00492-7.

Olsson, W. A., J. C. Lorenz, and S. P. Cooper, 2004, A mechanical model for multiply-oriented conjugate deformation bands: Journal of Structural Geology, v. 26, p. 325-338, doi:10.1016/s0191-8141(03)00101-9.

Parry, W. T., M. A. Chan, and B. Beitler, 2004, Chemical bleaching indicates episodes of fluid flow in deformation bands in sandstone: AAPG Bulletin, v. 88, p. 175-191, doi:10.1306/09090303034.

Paxton, S. T., J. O. Szabo, J. M. Ajdukiewicz, and R. E. Klimentidis 2002, Construction of an intergranular volume compaction curve for evaluating and predicting compaction and porosity loss in rigid-grain sandstone reservoirs: AAPG Bulletin, v. 86, p. 2047-2067, doi:10.1306/61EEDDFA-173E-11D7-8645000102C1865D.

Peisker, J., 2013, Vorhersagbarkeit der Reservoireigenschaften in Rotliegend Sandsteinen des Gebietes Dreilingen - NE Hannover, Deutschland: unpubl. Diplom thesis, Friedrich-Schiller-Universität Jena, Jena, Germany.

Perez, R. J., J. I. Chatellier, and R. H. Lander, 1999, Use of quartz cementation kinetic modeling to constrain burial histories. Examples from the Maracaibo Basin, Venezuela: Revista Latino Americana de Geoquimica Organica, v. 5, p. 39-46.

Pharaoh, T. C., M. Dusar, M. C. Geluk, F. Kockel, C. Krawczyk, P. Krzywiec, M. Scheck-Wenderoth, H. Thybo, O. V. Vejbæk, and J. D. Van Wees, 2010, Tectonic evolution, *in* H. Doornenbal, and A. Stevenson, eds., Petroleum Geological Atlas of the Southern Permian Basin Area: Houten, EAGE Publications BV, p. 25-57.

Pittmann, E. D., R. E. Larese, and M. T. Heald, 1992, Clay coats : Occurrence and relevance to preservation of porosity in sandstones, *in* D. W. Houseknecht, and E. D. Pittmann, eds., Origin, Diagenesis, and Petrophysics of Clay Minerals in Sandstones v. 47: Tulsa, Oklahoma, U.S.A., SEPM Special Publication, p. 241-256, doi:10.2110/pec.92.47.0241.

Platt, J. D., 1993, Controls on clay mineral distribution and chemistry in the early Permian Rotliegend of Germany: Clay Minerals, v. 28, p. 393-416, doi:10.1180/claymin.1993.028.3.05.

Ramm, M., 2000, Reservoir quality and its relationship to facies and provenance in Middle to Upper Jurassic sequences, northeastern North Sea: Clay Minerals, v. 35, p. 77-94, doi:10.1180/000985500546747.

Remy, R. R., 1994, Porosity reduction and major controls on diagenesis of Cretaceous-Paleocene volcaniclastic and arkosic sandstone, Middle Park Basin, Colorado: Journal of Sedimentary Research, v. A64, p. 797-806, doi:10.1306/D4267EC9-2B26-11D7-8648000102C1865D.

Rieke, H., 2001, Sedimentologie, Faziesarchtektur und Faziesentwicklung des kontinentalen Rotliegenden im Norddeutschen Becken: PhD thesis, University of Potsdam, Potsdam, 337 p.

Rimstidt, J. D., and H. L. Barnes, 1980, The kinetics of silica-water reactions: Geochimica et Cosmochimica Acta, v. 62, p. 1851-1863, doi:10.1016/0016-7037(80)90220-3.

Rogers, J. J. W., and W. B. Head, 1961, Relationships between porosity, median size, and sorting coefficients of synthetic sands: Journal of Sedimentary Petrology, v. 31, p. 467-470, doi:10.1306/74D70BA5-2B21-11D7-8648000102C1865D.

Roos, B. M., and B. J. Smits, 1983, Rotliegend and main Buntsandstein gas fields in Block K/13; a case history: Geologie en Mijnbouw. Netherlands Journal of Geosciences v. 62, p. 75-82.

Rotevatn, A., H. Fossen, J. Hesthammer, T. E. Aas, and J. A. Howell, 2007, Are relay ramps conduits for fluid flow? Structural analysis of a relay ramp in Arches National Park, Utah, *in* L. Lonergan, R. J. H. Jolly, K. Rawnsley, and D. J. Sanderson, eds.,

Fractured Reservoirs, v. 270, Geological Society, London, Special Publications, p. 55-71, doi:10.1144/gsl.sp.2007.270.01.04.

Rotevatn, A., and H. Fossen, 2011, Simulating the effect of subseismic fault tails and process zones in a siliciclastic reservoir analogue: Implications for aquifer support and trap definition: Marine and Petroleum Geology, v. 28, p. 1648-1662, doi:10.1016/j.marpetgeo.2011.07.005.

Ryan, P., and R. Reynolds, 1996, The origin and diagenesis of grain-coating serpentine-chlorite in Tuscaloosa Formation sandstone, US Gulf Coast: American Mineralogist, v. 81, p. 213-225, doi:10.2138/am-1996-1-226.

Saillet, E., and C. A. J. Wibberley, 2013, Permeability and flow impact of faults and deformation bands in high-porosity sand reservoirs: Southeast Basin, France, analog: AAPG Bulletin, v. 97, p. 437-464, doi:10.1306/09071211191.

Salem, A. M., S. Morad, L. F. Mato, and I. S. Al-Aasm, 2000, Diagenesis and reservoir-quality evolution of fluvial sandstones during progressive burial and uplift: Evidence from the Upper Jurassic Boipeba Member, Recôncavo Basin, Northeastern Brazil: AAPG Bulletin, v. 84, p. 1015-1040, doi:10.1306/A9673B9E-1738-11D7-8645000102C1865D.

Schöner, R., and R. Gaupp, 2005, Contrasting red bed diagenesis: the southern and northern margin of the Central European Basin: International Journal of Earth Sciences, v. 94, p. 897-916, doi:10.1007/s00531-005-0004-3.

Schöner, R., 2006, Comparison of Rotliegend sandstone diagenesis from the northern and southern margin of the North German Basin, and implications for the importance of organic maturation and migration: PhD thesis, Friedrich-Schiller-Universität Jena, Jena, 230 p.

Schueller, S., A. Braathen, H. Fossen, and J. Tveranger, 2013, Spatial distribution of deformation bands in damage zones of extensional faults in porous sandstones: Statistical analysis of field data: Journal of Structural Geology, v. 52, p. 148-162, doi:10.1016/j.jsg.2013.03.013.

Schultz, R. A., and H. Fossen, 2002, Displacement-length scaling in three dimensions: the importance of aspect ratio and application to deformation bands: Journal of Structural Geology, v. 24, p. 1389-1411, doi:10.1016/S0191-8141(01)00146-8.

Schultz, R. A., R. Soliva, H. Fossen, C. H. Okubo, and D. M. Reeves, 2008, Dependence of displacement–length scaling relations for fractures and deformation bands on the volumetric changes across them: Journal of Structural Geology, v. 30, p. 1405-1411, doi:10.1016/j.jsg.2008.08.001.

Schultz, R. A., C. H. Okubo, and H. Fossen, 2010, Porosity and grain size controls on compaction band formation in Jurassic Navajo Sandstone: Geophysical Research Letters, v. 37, p. 1-5, doi:10.1029/2010GL044909.

Schultz, R. A., C. Klimczak, H. Fossen, J. E. Olson, U. Exner, D. M. Reeves, and R. Soliva, 2013, Statistical tests of scaling relationships for geologic structures: Journal of Structural Geology, v. 48, p. 85-94, doi:10.1016/j.jsg.2012.12.005.

Schutjens, P. M. T. M., T. H. Hanssen, M. H. H. Hettema, J. Merour, P. de Bree, J. W. A. Coremans, and G. Helliesen, 2004, Compaction-induced porosity/permeability reduction in sandstone reservoirs: Data and model for elasticity-dominated deformation: SPE Reservoir Evaluation & Engineering, v. 7, p. 202-216, doi:10.2118/88441-PA.

Schwarzer, D., and R. Littke, 2007, Petroleum generation and migration in the 'Tight Gas' area of the German Rotliegend natural gas play: a basin modeling study: Petroleum Geoscience, v. 13, p. 37-62, doi:10.1144/1354-079306-703

Seemann, U., 1982, Depositional facies, diagenetic clay minerals and reservoir quality of Rotliegend sediments in the Southern Permian Basin (North Sea): A Review: Clay Minerals, v. 17, p. 55-67, doi:10.1180/claymin.1982.017.1.06.

Shipton, Z. K., J. P. Evans, and L. B. Thompson, 2005, The geometry and thickness of deformation-band fault core and its influence on sealing characteristics of deformation-band fault zones, *in* R. Sorkhabi, and Y. Tsuji, eds., Faults, Fluid Flow, and Petroleum Traps AAPG Memoir, v. 85, p. 181-195, doi:10.1306/1033723m853135.

Sindern, S., H. Stanjek, C. Hilgers, and Y. Etoundi, 2007, Short-term hydrothermal effects on the 'crystallinities' of illite and chlorite in the footwall of the Aachen-Faille du midi thrust fault – first results of the RWTH-1 drilling project: Clays and Clay Minerals, v. 55, p. 200-212, doi:10.1346/ccmn.2007.0550209.

Smith, W. O., P. D. Foote, and P. F. Busang, 1930, Capillary retention of liquids in assemblages of homogeneous spheres: Physical Review, v. 36, p. 524-530, doi:10.1103/PhysRev.36.524.

Soliva, R., R. A. Schultz, G. Ballas, A. Taboada, C. Wibberley, E. Saillet, and A. Benedicto, 2013, A model of strain localization in porous sandstone as a function of tectonic setting, burial and material properties; new insight from Provence (southern France): Journal of Structural Geology, v. 49, p. 50-63, doi:10.1016/j.jsg.2012.11.011.

Sternlof, K. R., 2006, Structural geology, propagation mechanics and hydraulic effects of compaction bands in sandstone: PhD thesis, Stanford University, Stanford, 214 p.

Storvoll, V., K. Bjørlykke, D. Karlsen, and G. Saigal, 2002, Porosity preservation in reservoir sandstones due to grain-coating illite: a study of the Jurassic Garn Formation from the Kristin and Lavrans fields, offshore Mid-Norway: Marine and Petroleum Geology, v. 19, p. 767-781, doi:10.1016/S0264-8172(02)00035-1.

Takeno, N., 2005, Atlas of Eh-pH diagrams: Intercomparison of thermodynamic databases, Geological Survey of Japan, p. 287.

Taylor, T., R. Stancliffe, C. Macaulay, and L. Hathon, 2004, High temperature quartz cementation and the timing of hydrocarbon accumulation in the Jurassic Norphlet sandstone, offshore Gulf of Mexico, USA, *in* J. M. Cubitt, W. A. England, and S. Larter, eds., Understanding Petroleum Reservoirs: Towards an Integrated Reservoir Engineering and Geochemical Approach, v. 237: London, Geological Society, p. 257-278, doi:10.1144/GSL.SP.2004.237.01.15.

Taylor, T. R., M. R. Giles, L. A. Hathon, T. N. Diggs, N. R. Braunsdorf, G. V. Birbiglia, M. G. Kittridge, C. I. Macaulay, and I. S. Espejo, 2010, Sandstone diagenesis and reservoir quality prediction: Models, myths, and reality: AAPG Bulletin, v. 94, p. 1093-1132, doi:10.1306/04211009123.

Taylor, T. R., M. G. Kittridge, P. Winefield, L. T. Bryndzia, and L. M. Bonnell, 2015, Reservoir quality and rock properties modeling – Triassic and Jurassic sandstones, greater Shearwater area, UK Central North Sea: Marine and Petroleum Geology, v. 65, p. 1-21, doi:10.1016/j.marpetgeo.2015.03.020.

Tembe, S., P. Baud, and T.-f. Wong, 2008, Stress conditions for the propagation of discrete compaction bands in porous sandstone: Journal of Geophysical Research, v. 113, p. 1-16, doi:10.1029/2007jb005439.

Torabi, A., 2014, Cataclastic bands in immature and poorly lithified sandstone, examples from Corsica, France: Tectonophysics, v. 630, p. 91-102, doi:10.1016/j.tecto.2014.05.014.

Tucker, M. E., 2001, Sedimentary Petrology: Oxford, Blackwell Science Ltd, 272 p.

Tueckmantel, C., Q. J. Fisher, C. A. Grattoni, and A. C. Aplin, 2012, Single- and two-phase fluid flow properties of cataclastic fault rocks in porous sandstone: Marine and Petroleum Geology, v. 29, p. 129-142, doi:10.1016/j.marpetgeo.2011.07.009.

Turner, P., S. D. Burley, D. Rey, and J. Prosser, 1995, Burial history of the Penrith Sandstone (Lower Permian) deduced from the combined study of fluid inclusion and palaeomagnetic data: Geological Society, London, Special Publications, v. 98, p. 43-78, doi:10.1144/gsl.sp.1995.098.01.04.

Uffmann, A. K., and R. Littke, 2011, 3D petroleum systems modelling of the North German Basin: first break, v. 29, p. 49-63, doi:10.3997/1365-2397.2011016

Underhill, J. R., R. A. Gayer, N. H. Woodcock, R. Donnelly, E. J. Jolley, and I. G. Stimpson, 1988, The Dent Fault System, northern England-reinterpreted as a major oblique-slip fault zone: Journal of the Geological Society, v. 145, p. 303-316, doi:10.1144/gsjgs.145.2.0303.

Vackiner, A. A., 2011, Sedimentary facies reconstruction and kinematic restoration of an Upper Permian tight gas field, north-western Germany: PhD thesis, RWTH Aachen University, Aachen, 144 p.

Vajdova, V., P. Baud, and T.-f. Wong, 2004, Permeability evolution during localized deformation in Bentheim sandstone: Journal of Geophysical Research, v. 109, p. 1-15, doi:10.1029/2003JB002942.

van den Belt, F. J. G., and F. F. N. van Hulten, 2011, Sedimentary architecture and palaeogeography of Lower Slochteren aeolian cycles from the Rotliegend desert-lake margin (Permian), The Markham area, southern North Sea, *in* J. Grötsch, and R. Gaupp, eds., The Permian Rotliegend of the Netherlands, v. 98: Tulsa, Oklahoma, U.S.A., Society for Sedimentary Geology, doi:doi:10.2110/pec.11.98.0161.

van Houten, F. B., 1973, Origin of red beds: A review - 1961-1972: Annual Review of Earth and Planetary Sciences, v. 1, p. 39-61, doi:10.1146/annurev.ea.01.050173.000351.

Verweij, H., 2003, Fluid flow systems analysis on geological timescales in onshore and offshore Netherlands. With special reference to the Broad Fourteens Basin: PhD thesis, Vrije Universiteit Amsterdam, 290 p.

Walderhaug, O., 1994, Precipitation rates for quartz cement in sandstones determined by fluid-inclusion microthermometry and temperature-history modeling: Journal of Sedimentary Research, v. A64, p. 324-333, doi:10.2110/jsr.64.324.

Walderhaug, O., 1996, Kinetic modelling of quartz cementation and porosity loss in deeply buried sandstone reservoirs: AAPG Bulletin, v. 80, p. 731-745, doi:10.1306/64ED88A4-1724-11D7-8645000102C1865D.

Walderhaug, O., 2000, Modeling quartz cementation and porosity in middle Jurassic Brent Group sandstones of the Kvitebjorn Field, Northern North Sea: AAPG Bulletin, v. 84, p. 1325-1339, doi:10.1306/A9673E96-1738-11D7-8645000102C1865D.

Walderhaug, O., R. H. Lander, P. A. Bjørkum, H. Oelkers, K. Bjørlykke, and P. H. Nadeau, 2000, Modelling quartz cementation and porosity in reservoir sandstones: examples from the Norwegian continental shelf, *in* R. H. Worden, and S. Morad, eds., Quartz Cementation in Sandstones, v. 29: Oxford, Blackwell Science Ltd., p. 39-49, doi:10.1002/9781444304237.ch3.

Walker, T. R., 1979, Red color in dune sands, *in* E. D. McKee, ed., A study of global sand seas: Geological Survey Professional Paper: Washington, D. C., U.S. Government Printing Office, p. 61-81.

Waugh, B., 1970a, Formation of quartz overgrowths in the Penrith Sandstone (Lower Permian) of Northwest England as revealed by scanning electron microscopy: Sedimentology, v. 14, p. 309-320, doi:10.1111/j.1365-3091.1970.tb00197.x.

Waugh, B., 1970b, Petrology, provenance and silica diagenesis of the Penrith Sandstone (Lower Permian) of Northwest England: Journal of Sedimentary Petrology, v. 40, p. 1226-1240, doi:10.1306/74D72171-2B21-11D7-8648000102C1865D.

Waugh, B., and L. C. Macchi, 1988, Poroperm Excursion Guide No. 2: The Permo Triassic rocks of Cumbria - Supplement to Excursion Guide 2: Chester, England, Poroperm-Geochem Ltd.

Wennberg, O. P., G. Casini, A. Jahanpanah, F. Lapponi, J. Ineson, B. G. Wall, and P. Gillespie, 2013, Deformation bands in chalk, examples from the Shetland Group of the Oseberg Field, North Sea, Norway: Journal of Structural Geology, v. 56, p. 103-117, doi:10.1016/j.jsg.2013.09.005.

Williams, G. J., 2011, Data mining with R and Rattle: The art of excavating data for knowledge discovery: New York, Springer.

Wilson, M. D., 1992, Inherited grain-rimming clays in sandstones from eolian and shelf environments : Their origin and control on reservoir properties, *in* D. W. Houseknecht, and E. D. Pittman, eds., Origin, Diagenesis, and Petrophysics of Clay Minerals in Sandstone, v. 47: Tulsa, Oklahoma, U.S.A., SEPM Special Publications, p. 282.

Winters, C. G., 1989, An integrated sedimentological study of the cored interval from Well K/10-14, Rijswijk, The Nederlands, Core Laboratories.

Wong, T.-f., C. David, and W. Zhu, 1997, The transition from brittle faulting to cataclastic flow in porous sandstones: Mechanical deformation: Journal of Geophysical Research, v. 102, p. 3009-3025, doi:10.1029/96jb03281.

Woodcock, N. H., and C. Schubert, 1994, Continental strike-slip tectonics, *in* P. Hancock, ed., Continental Deformation: Oxford, Pergamon Press Ltd, p. 251-263.

Worden, R. H., and S. Morad, 2002, Clay minerals in sandstones: controls on formation, distribution and evolution, *in* R. H. Worden, and S. Morad, eds., Clay Mineral Cements in Sandstones: Oxford, Blackwell Science Ltd., p. 3-42, doi:10.1002/9781444304336.ch1.

Wygrala, B. P., 1989, Integrated study of an oil field in the southern Po basin, northern Italy, Kernforschungsanlage Jülich reports, p. 217.

Zhu, C., and T.-f. Wong, 1997, The transition from brittle faulting to cataclastic flow: Permeability evolution: Journal of Geophysical Research, v. 102, p. 3027-3041, doi:10.1029/96jb03282.

Ziegler, P. A., 1990, Geological atlas of Western and Central Europe: The Hague, Shell Internationale Petroleum Maatschapij B.V., 239 p.

Zwingmann, H., N. Clauer, and R. Gaupp, 1998, Timing of fluid flow in a sandstone reservoir of the north german Rotliegend (Permian) by K-Ar dating of related hydrothermal illite, *in* J. Parnell, ed., Dating and Duration of Fluid Flow and Fluid-Rock Interaction, v. 144, Geological Society, London, Special Publications,, p. 91-106.

Zwingmann, H., N. Clauer, and R. Gaupp, 1999, Structure-related geochemical (REE) and isotopic (K-Ar, Rb-Sr, d18O) characteristics of clay minerals from Rotliegend sandstone reservoirs (Permian, northern Germany): Geochimica et Cosmochimica Acta, v. 63, p. 2805-2823, doi:10.1016/S0016-7037(99)00198-2.

8 Appendices

Table A.1: Compilation of point count data and sample information of well A-E.

Sample ID	Well	Depositional Environment	Quartz	Feldspars	Rock fragments	Accessories	Iron oxides	Depositional matrix	Quartz cement	Feldspar cement	Carbonate cement	Sulfate cement	Illite cement	Illite replacement	Chlorite cement	Chlorite replacement	Intergranular porosity	Secondary porosity	Intergranular volume (%)	Coat coverage (%)	Plug porosity (%)	Plug permeability (mD)
A1	A	Dry Sandflat	55,33	12,00	3,00	0,00	0,00	3,00	11,00	1,33	6,00	0,00	0,00	2,00	0,00	0,00	5,33	1,00	23,67	70,50	8,15	8,70
A2	A	Dune	61,67	10,00	1,67	0,00	0,33	5,00	3,67	0,00	1,33	12,00	0,00	2,33	1,67	0,33	0,00	0,00	18,66	80,50	2,01	0,07
A3	A	Damp	53,00	11,00	3,33	0,00	0,00	2,67	19,33	1,67	1,00	0,33	0,00	2,67	0,33	0,00	2,67	2,00	25,33	79,92	6,65	1,06
A4	A	Dry Sandflat	56,33	11,67	1,67	0,00	0,00	1,67	9,00	2,67	2,00	0,33	0,00	1,00	2,67	0,00	9,67	1,33	26,33	88,84	12,80	43,50
A5	A	Dry Sandflat	60,67	8,67	1,00	0,33	0,33	1,67	6,67	1,00	9,66	1,66	0,67	2,00	3,67	0,00	1,67	0,33	24,99	86,16	5,91	0,17
A6	A	Dry Sandflat	60,67	9,00	4,67	0,33	0,00	1,67	5,00	1,00	3,67	0,00	0,00	1,67	4,67	0,00	7,00	0,67	21,33	92,72	11,80	43,50
A7	A	Dry Sandflat	63,00	12,33	2,00	0,00	0,00	1,33	4,00	1,00	4,67	1,00	0,00	2,67	4,00	0,00	3,67	0,33	18,33	87,22	8,68	4,20
A8	A	Dry Sandflat	61,33	7,00	2,67	0,33	0,00	1,00	9,33	1,00	9,33	1,00	0,00	2,00	2,67	0,33	1,33	0,67	24,67	68,08	5,17	0,10
A9	A	Dry Sandflat	69,33	9,67	0,67	0,00	0,00	0,67	1,67	0,67	1,00	0,33	0,00	1,67	6,00	0,00	7,00	1,33	16,67	97,22	20,60	n.a.
B1	B	Dune	51,00	9,67	5,67	0,33	1,33	2,33	12,67	0,67	7,00	7,67	0,00	1,00	0,00	0,00	0,67	0,00	28,67	48,50	2,10	0,06
B2	B	Dry Sandflat	63,99	4,33	1,67	0,00	0,33	2,00	10,67	1,33	11,67	3,00	0,00	1,00	0,00	0,00	0,00	0,00	26,66	35,12	1,70	0,03
B3	B	Dune	55,32	6,00	1,33	0,00	1,33	1,33	13,33	1,00	10,33	9,67	0,00	0,33	0,00	0,00	0,00	0,00	34,33	20,18	2,30	0,04
B4	B	Dune	56,00	6,33	1,00	0,00	1,00	1,67	13,67	1,00	10,33	8,00	0,00	0,33	0,00	0,00	0,33	0,33	33,33	13,80	2,30	0,05
B5	B	Dry Sandflat	57,33	9,00	3,30	0,00	1,00	1,33	11,67	0,67	7,67	2,33	0,00	0,67	0,00	0,33	4,33	0,33	26,66	9,30	8,60	0,66
B6	B	Dry Sandflat	53,67	8,00	5,67	0,67	0,33	1,67	13,00	0,67	1,67	4,00	0,00	1,00	0,00	0,00	9,00	0,00	28,34	13,70	10,80	11,70
B7	B	Dry Sandflat	63,33	6,33	2,67	0,00	0,33	0,67	7,67	2,67	0,33	5,33	0,00	1,00	0,00	0,00	9,33	0,33	25,33	11,80	11,10	26,50
B8	B	Dry Sandflat	67,66	6,00	3,67	0,33	0,00	0,33	12,00	1,33	0,67	1,67	0,00	1,30	0,00	0,00	4,00	0,00	19,67	10,60	10,30	11,10
B9	B	Damp	60,00	4,33	2,67	1,00	0,67	5,67	5,00	0,30	1,00	14,66	3,67	1,00	0,00	0,00	0,00	0,00	24,62	22,90	2,00	0,02
B10	B	Dry Sandflat	61,67	6,33	5,67	0,00	0,67	1,67	12,33	2,33	4,67	2,33	0,33	0,33	0,00	0,00	1,67	0,00	23,67	18,60	5,20	0,09
C1	C	Dune	65,33	8,00	5,67	0,67	0,00	1,33	4,00	0,33	3,00	0,33	1,00	2,33	6,33	0,00	0,67	1,00	15,66	92,42	4,50	0,04
C2	C	Dry Sandflat	67,33	6,67	3,67	0,00	0,00	1,00	8,67	0,00	7,67	2,33	0,33	1,00	0,33	0,33	0,33	0,33	19,66	83,36	3,50	0,04
C3	C	Dry Sandflat	64,00	8,00	2,00	0,00	0,00	0,00	11,00	0,00	5,00	1,00	2,33	1,33	3,33	0,00	1,67	0,00	24,33	85,36	4,20	0,03
C4	C	Dune	56,33	8,67	7,67	0,00	0,33	0,33	6,33	1,67	4,00	4,33	0,00	1,00	4,33	0,33	4,00	0,33	24,66	97,26	5,80	0,09
C5	C	Dune	62,33	9,67	2,00	0,33	0,00	1,00	5,00	1,33	2,33	5,67	0,33	2,00	2,33	0,33	5,00	0,33	22,00	89,36	8,90	1,46
C6	C	Dune	54,67	10,00	4,00	0,00	0,00	0,33	8,33	1,00	1,67	12,33	0,33	0,67	2,33	1,00	2,67	0,67	28,66	79,10	5,90	0,25
C7	C	Dry Sandflat	58,33	12,00	4,67	0,00	0,00	0,67	5,67	0,67	2,33	5,33	0,00	0,67	4,67	0,00	0,66	0,33	19,32	89,42	8,20	2,63
C8	C	Damp	59,00	6,67	3,67	0,33	0,00	0,00	4,00	0,67	1,67	0,00	13,00	1,00	7,67	1,00	0,00	1,67	27,00	87,84	0,90	0,01
D1	D	Dune	60,00	10,33	2,33	1,33	0,33	0,33	11,33	1,33	4,00	5,67	0,33	0,33	0,00	0,00	2,33	0,00	25,00	32,40	4,70	0,34
D2	D	Dune	49,00	11,33	4,00	0,67	0,33	0,33	12,67	0,33	8,00	7,00	0,00	1,33	0,00	0,33	4,00	0,67	32,00	41,20	6,40	2,37
D3	D	Dune	58,00	11,00	1,67	0,33	0,00	0,33	9,67	1,67	2,67	6,67	0,00	1,00	0,00	0,33	4,33	2,33	25,00	42,90	8,00	13,40
D4	D	Dry Sandflat	61,67	9,00	1,00	0,00	0,33	0,00	17,67	1,00	5,99	1,04	0,00	1,33	0,00	0,00	0,33	0,67	26,03	8,80	4,30	0,23
D5	D	Dry Sandflat	58,00	9,00	3,33	0,33	0,33	0,33	7,33	0,67	16,33	2,66	0,00	0,67	0,00	0,67	0,00	0,33	26,99	5,40	3,80	0,07

Table A.1 (continued)

Sample ID	Well	Depositional Environment	Quartz	Feldspars	Rock fragments	Accessories	Iron oxides	Depositional matrix	Quartz cement	Feldspar cement	Carbonate cement	Sulfate cement	Illite cement	Illite replacement	Chlorite cement	Chlorite replacement	Intergranular porosity	Secondary porosity	Intergranular volume (%)	Coat coverage (%)	Plug porosity (%)	Plug permeability (mD)
D6	D	Damp	59,00	12,33	1,00	0,67	2,67	0,67	16,33	0,67	1,33	2,00	0,00	0,67	0,00	0,00	2,00	0,67	22,33	68,80	6,50	0,29
D7	D	Dry Sandflat	56,67	10,00	5,00	0,00	3,67	0,00	18,33	0,33	1,67	0,67	0,00	0,67	0,00	0,67	0,33	1,67	21,33	28,60	6,10	0,15
D8	D	Dry Sandflat	57,67	15,00	2,00	0,00	1,00	0,33	7,67	0,33	11,67	2,66	0,00	1,00	0,00	0,33	0,00	0,33	22,33	13,26	3,00	0,06
D9	D	Dry Sandflat	50,67	9,33	1,67	0,33	0,67	0,33	11,33	2,00	9,00	14,67	0,00	0,00	0,00	0,00	0,00	0,00	37,00	19,92	1,90	n.a.
D10	D	Dry Sandflat	55,67	7,67	1,67	0,00	0,33	0,67	6,67	0,00	6,33	20,00	0,00	0,67	0,00	0,33	0,00	0,00	33,00	15,30	1,10	0,03
E1	E	Lake Margin	48,00	9,00	2,00	0,33	0,33	2,00	8,00	0,33	12,00	17,00	0,00	1,00	0,00	0,00	0,00	0,00	37,33	65,14	0,93	35,00
E2	E	Dry Sandflat	63,00	8,67	3,67	0,00	0,00	0,00	9,00	2,00	3,67	1,67	0,00	0,00	0,00	0,00	7,00	1,33	23,33	30,20	9,25	5,10
E3	E	Dry Sandflat	64,33	7,00	1,33	0,00	0,00	1,00	12,00	2,33	8,33	1,33	0,00	0,67	0,00	0,33	0,67	0,33	24,66	66,18	n.a.	n.a.
E4	E	Dry Sandflat	63,67	11,00	3,00	0,00	0,00	1,00	5,67	1,33	1,00	1,33	0,00	0,33	0,00	0,33	9,33	2,00	18,66	81,26	11,70	22,30
E5	E	Dry Sandflat	62,00	10,33	2,00	0,67	0,00	0,33	5,33	1,00	6,66	1,00	0,33	0,00	0,00	0,33	8,33	1,67	22,66	84,86	13,90	41,80
E6	E	Dry Sandflat	57,33	8,00	1,00	0,33	0,00	1,00	22,67	1,67	4,33	0,00	0,00	0,67	0,00	0,00	2,33	0,67	30,99	70,46	5,80	0,13
E7	E	Dry Sandflat	55,67	7,67	1,00	0,33	0,33	0,66	9,33	2,00	10,00	3,67	0,67	1,00	0,00	0,00	7,33	0,33	33,00	90,30	6,90	0,53
E8	E	Dry Sandflat	54,67	5,00	1,67	0,00	0,00	0,00	8,00	2,00	13,33	15,33	0,00	0,00	0,00	0,00	0,00	0,00	38,66	8,30	1,80	0,05
E9	E	Dry Sandflat	59,67	7,66	1,33	0,33	0,00	1,00	3,67	1,00	13,67	2,00	3,00	2,67	0,00	0,00	4,00	0,00	27,33	41,20	5,30	0,26
E10	E	Dry Sandflat	57,33	12,00	1,67	0,00	0,00	0,67	3,67	1,00	1,33	4,67	0,00	1,67	0,00	0,33	12,66	1,00	23,33	91,32	15,90	122,00
E11	E	Dry Sandflat	58,00	7,67	3,67	0,00	0,00	0,67	10,00	0,67	3,33	2,00	1,00	1,00	1,33	0,00	8,66	1,33	26,99	87,48	10,60	15,00
E12	E	Dry Sandflat	61,00	12,33	0,00	0,33	0,00	0,00	5,67	0,67	14,66	3,00	0,33	0,66	0,00	0,33	0,00	1,00	24,32	48,60	4,20	0,05
E13	E	Dry Sandflat	53,67	6,33	2,33	0,00	0,00	0,67	11,33	0,33	9,67	11,00	1,00	1,33	0,00	0,00	1,33	1,00	34,66	34,60	2,80	0,08

Table A.2: Compilation of point count data and sample information of well F-J.

Sample ID	Well	Depositional Environment	Quartz	Feldspars	Rock fragments	Accessories	Depositional matrix	Quartz cement	Feldspar cement	Carbonate cement	Sulfate cement	Illite cement	Kaolinite cement	Intergranular porosity	Intergranular volume (%)	Coat coverage (%)	Plug porosity (%)	Plug permeability (mD)
F1	F	Damp sandflat	60,30	0,00	2,51	0,50	0,25	1,67	0,00	12,56	0,50	2,51	4,19	15,00	36,69	84,72	17,20	29,00
F2	F	Damp sandflat	65,90	0,00	2,53	0,25	0,00	1,69	0,00	7,60	0,25	2,53	4,22	15,00	31,31	76,30	16,20	24,00
F3	F	Damp sandflat	67,59	0,89	2,67	0,53	0,00	1,78	0,00	8,89	0,53	2,67	4,45	10,00	28,32	78,04	12,00	1,40
F4	F	Damp sandflat	55,60	0,25	2,53	0,51	0,00	1,68	0,00	16,85	0,00	1,68	5,90	15,00	41,11	85,08	18,70	107,00
F5	F	Damp sandflat	67,87	0,26	2,64	0,79	3,53	0,26	0,00	7,05	0,53	1,76	5,29	10,00	28,42	74,08	10,60	1,70
F6	F	Dry sandflat	52,17	0,24	2,37	0,71	0,00	2,37	0,00	11,86	0,00	2,37	7,91	20,00	44,51	80,58	20,20	179,00
F7	F	Damp sandflat	61,97	0,25	2,51	0,75	0,00	0,25	0,00	10,89	0,00	3,35	5,02	15,00	34,51	91,26	15,00	6,70
F8	F	Damp sandflat	58,01	0,27	3,63	0,82	4,53	0,27	0,00	10,88	0,00	4,53	9,06	8,00	37,28	66,57	14,10	4,30
F9	F	Dry sandflat	63,23	0,24	3,20	0,24	0,00	0,24	0,00	8,80	0,24	2,40	2,40	19,00	33,09	93,38	22,10	239,00
F10	F	Damp sandflat	63,67	0,26	3,49	0,52	0,00	0,87	0,00	6,98	0,00	6,11	6,11	12,00	32,06	83,48	19,20	70,00
G1	G	Reworked	64,42	0,99	0,99	0,30	0,00	0,00	0,00	14,87	0,59	0,00	4,96	12,88	33,30	43,90	18,60	31,00
G2	G	Dune	61,39	1,98	0,99	0,99	0,00	0,99	0,00	16,83	1,98	0,00	4,95	9,90	34,65	34,30	20,40	85,00
G3	G	Dune	35,89	4,99	5,98	1,00	0,00	0,00	0,00	16,95	0,30	1,99	14,96	17,95	52,14	19,20	21,30	225,00
G4	G	Low-energy fluvial	64,61	0,30	5,96	0,30	0,00	1,99	0,00	12,92	0,00	4,97	4,97	3,98	28,83	19,10	11,00	3,50
G5	G	Wet sandflat	59,64	3,98	7,95	0,99	0,00	1,99	0,30	11,93	0,30	2,98	3,98	5,96	27,44	28,90	8,90	49,00
G6	G	Dune	19,88	0,99	1,99	0,30	49,70	0,30	0,00	2,98	0,00	0,00	16,90	6,96	76,84	44,90		
G7	G	Dune	N/A	N/A	N/A	N/A	N/A	N/A	N/A	N/A	N/A	N/A	N/A	N/A	N/A	19,30		
G8	G	Dune	4,91	0,98	0,98	0,29	73,60	0,00	0,98	3,93	0,29	0,29	2,94	10,79	92,84	35,60	15,20	1,50
G9	G	Damp sandflat	61,63	2,98	1,99	1,99	0,99	0,30	0,00	21,87	0,30	0,00	5,96	1,99	31,41	37,52	13,40	2,60
G10	G	Lake margin	70,00	2,00	3,00	1,00	1,00	1,00	0,00	17,00	1,00	1,00	3,00	0,00	24,00	30,80	6,30	8,70
G11	G	Dune	48,00	5,00	6,00	2,00	1,00	1,00	0,00	5,00	1,00	1,00	17,00	13,00	39,00	31,30	22,60	734,00
G12	G	Dry sandflat	42,00	3,00	2,00	3,00	7,00	2,00	0,00	8,00	1,00	2,00	15,00	15,00	50,00	34,60	25,10	530,00
G13	G	Dry sandflat	75,77	2,99	1,00	1,99	1,99	1,00	0,30	4,99	1,00	1,99	4,99	1,99	18,25	29,90	21,00	33,00
G14	G	Dry sandflat	55,67	1,99	4,97	2,98	0,30	0,00	0,00	6,96	0,30	1,99	4,97	19,88	34,39	33,20	20,90	160,00
G15	G	Damp sandflat	58,00	1,00	1,00	4,00	5,00	1,00	0,00	18,00	0,00	1,00	8,00	3,00	36,00	23,62	18,30	55,00
G16	G	Damp sandflat	32,90	4,99	2,99	3,29	3,99	0,00	0,00	16,95	1,00	1,00	17,95	14,96	55,83	40,10	20,70	115,00
G17	G	Dry sandflat	43,74	1,99	4,97	0,99	0,99	0,99	0,00	15,90	5,27	0,30	9,94	14,91	48,31	34,74	17,50	65,00
G18	G	Dry sandflat	59,82	1,99	1,99	3,29	4,99	0,00	0,00	5,98	0,00	2,99	11,96	6,98	32,90	36,78	19,80	71,00
G19	G	Dry sandflat	54,05	6,01	9,01	3,00	0,00	0,30	0,30	2,00	0,30	2,00	7,01	16,02	27,93	39,30	21,50	630,00
G20	G	Dry sandflat	48,90	1,00	0,90	2,00	0,30	0,00	0,00	3,99	0,00	2,99	19,96	19,96	47,21	37,10	17,80	4,60
G21	G	Dry sandflat	45,82	1,00	15,44	2,29	0,30	0,30	0,00	5,98	0,00	1,99	11,95	14,94	35,46	22,30	24,60	508,00
G22	G	Dry sandflat	54,84	7,98	3,99	1,99	0,00	0,00	0,30	2,99	0,00	1,00	4,99	21,93	31,21	27,80	23,10	827,00

Table A.2 (continued)

Sample ID	Well	Depositional Environment	Quartz	Feldspars	Rock fragments	Accessories	Depositional matrix	Quartz cement	Feldspar cement	Carbonate cement	Sulfate cement	Illite cement	Kaolinite cement	Intergranular porosity	Intergranular volume (%)	Coat coverage (%)	Plug porosity (%)	Plug permeability (mD)
G23	G	Low-energy fluvial	65,80	2,99	3,99	1,00	0,00	1,99	0,00	2,99	0,30	2,99	5,98	11,96	26,22	31,20	20,30	82,00
G24	G	Dry sandflat	62,62	1,99	4,97	1,99	8,95	0,30	0,00	2,98	0,30	0,99	7,95	6,96	28,43	35,70	20,40	12,00
G25	G	Low-energy fluvial	61,81	1,99	1,99	1,00	1,00	0,00	0,00	3,99	4,29	2,99	3,99	16,95	33,20	34,42	22,90	359,00
G26	G	Low-energy fluvial	N/A	N/A	N/A	N/A	N/A	N/A	N/A	N/A	N/A	N/A	N/A	N/A	N/A	62,26		
G27	G	Low-energy fluvial	N/A	N/A	N/A	N/A	N/A	N/A	N/A	N/A	N/A	N/A	N/A	N/A	N/A	56,88		
G28	G	Dune	66,60	1,99	2,98	2,29	2,98	0,30	0,00	6,96	0,00	1,99	11,93	1,99	26,14	35,60	18,20	10,00
G29	G	Dune	N/A	N/A	N/A	N/A	N/A	N/A	N/A	N/A	N/A	N/A	N/A	N/A	N/A	32,90		
G30	G	Dry sandflat	53,84	0,30	4,99	4,99	0,00	1,00	0,00	9,97	0,00	1,99	7,98	14,96	35,89	39,60	10,80	1,10
G31	G	Low-energy fluvial	48,56	4,96	1,98	1,29	0,00	0,00	0,00	3,96	3,27	2,97	17,84	14,87	43,21	42,00	22,70	1535,00
G32	G	Low-energy fluvial	55,83	1,00	1,99	2,99	9,97	0,30	0,00	5,98	1,00	1,00	4,99	14,96	38,19	51,00	20,90	40,00
G33	G	Dune	56,00	4,00	5,00	2,00	7,00	2,00	0,00	1,00	0,00	3,00	13,00	7,00	33,00	56,20	17,50	68,00
G34	G	Dry sandflat	61,00	1,00	6,00	2,00	3,00	2,00	0,00	8,00	1,00	1,00	7,00	8,00	30,00	48,60	18,10	79,00
G35	G	Dry sandflat	73,05	1,97	3,95	2,27	7,90	2,96	0,00	2,96	0,00	0,99	1,97	1,97	18,76	53,00		
H1	H	Low-energy fluvial	56,78	0,30	4,98	1,99	1,99	1,00	0,00	24,90	0,00	1,00	3,98	3,08	35,95	75,48	9,00	0,33
H2	H	Low-energy fluvial	58,42	1,98	9,90	0,99	2,97	1,98	0,00	5,94	0,00	0,99	6,93	9,89	28,70	70,74	18,40	46,00
H3	H	Low-energy fluvial	64,00	0,00	7,00	1,00	10,00	4,00	0,00	9,00	0,00	2,00	3,00	0,00	28,00	38,60	2,70	0,13
H4	H	Dune	56,83	0,30	6,98	1,00	3,99	1,00	0,00	8,97	0,00	1,99	4,99	13,96	34,90	83,58	20,70	145,00
H5	H	Dry sandflat	62,62	0,00	5,96	0,30	2,98	0,99	0,00	5,96	1,99	0,30	3,98	14,91	31,11	82,68	14,60	38,00
H6	H	Low-energy fluvial	62,00	3,00	5,00	1,00	6,00	1,00	0,00	9,00	0,00	1,00	5,00	7,00	29,00	83,36	16,50	27,00
H7	H	Dry sandflat	55,67	1,99	0,99	1,29	11,93	2,98	0,00	9,94	0,99	0,99	2,98	9,94	40,06	63,20	15,80	306,00
H8	H	Damp sandflat	62,25	0,00	0,99	0,59	1,98	0,99	0,00	16,80	3,26	0,30	2,96	9,88	36,17	81,40	12,60	9,10
H9	H	Damp sandflat	50,00	4,00	7,00	1,00	4,00	1,00	0,00	10,00	0,00	2,00	5,00	16,00	38,00	72,46	21,90	203,00
H10	H	Damp sandflat	62,62	0,99	11,93	0,30	2,98	1,99	0,00	9,94	0,00	0,30	2,98	5,96	24,16	38,98	12,30	0,36
H11	H	Damp sandflat	63,62	0,00	1,99	0,30	6,96	0,99	0,00	4,97	1,99	0,30	0,99	17,89	34,10	68,94	15,20	21,00
H12	H	Damp sandflat	60,00	3,00	7,00	1,00	4,00	1,00	0,00	8,00	0,00	2,00	6,00	8,00	29,00	62,90	14,50	7,70
H13	H	Wet sandflat	44,60	0,00	3,96	9,91	29,73	0,30	0,00	8,92	0,00	0,30	1,98	0,30	41,53	40,46	3,30	70,00
H14	H	Lake margin	62,62	0,99	3,98	0,99	4,97	0,30	0,00	5,96	1,29	0,99	2,98	14,91	31,41	64,22	19,80	136,00
H15	H	Dry sandflat	58,00	2,00	6,00	1,00	4,00	1,00	0,00	11,00	0,00	2,00	7,00	8,00	33,00	59,08	14,30	2,70
H16	H	Dune	54,84	1,00	8,97	0,30	5,98	1,99	0,00	2,99	0,00	1,00	6,98	15,95	34,90	56,30	22,80	
H17	H	Dry sandflat	68,38	1,98	1,98	0,30	9,91	0,30	0,00	9,91	0,00	0,00	0,30	6,94	27,35	45,00	18,10	34,00
H18	H	Dry sandflat	62,81	3,99	1,99	1,00	4,99	0,00	0,00	4,99	0,30	1,99	2,99	14,96	30,21	78,06	17,10	33,00
H19	H	Dry sandflat	57,00	0,00	5,00	1,00	3,00	1,00	0,00	9,00	0,00	2,00	10,00	12,00	37,00	66,40	17,40	22,00

Table A.2 (continued)

Sample ID	Well	Depositional Environment	Quartz	Feldspars	Rock fragments	Accessories	Depositional matrix	Quartz cement	Feldspar cement	Carbonate cement	Sulfate cement	Illite cement	Kaolinite cement	Intergranular porosity	Intergranular volume (%)	Coat coverage (%)	Plug porosity (%)	Plug permeability (mD)
H20	H	Damp sandflat	60,00	0,00	6,00	1,00	1,00	1,00	0,00	4,00	0,00	1,00	10,00	16,00	33,00	69,78	20,90	61,00
H21	H	Low-energy fluvial	57,65	0,30	7,95	0,99	0,30	0,00	0,00	3,98	1,99	0,99	5,96	19,88	33,10	82,76	26,30	1530,00
H22	H	Low-energy fluvial	76,54	0,30	2,98	1,29	4,97	0,00	0,00	4,97	3,98	0,99	1,99	1,99	18,89	58,00	22,00	11,00
H23	H	Low-energy fluvial	59,00	3,00	8,00	1,00	4,00	1,00	0,00	6,00	0,00	2,00	6,00	10,00	29,00	78,62	17,10	1,00
H24	H	Low-energy fluvial	61,88	0,00	3,99	0,30	14,97	0,00	0,00	11,98	1,30	0,30	0,30	4,99	33,83	38,10	13,30	0,51
H25	H	Damp sandflat	62,62	0,30	4,97	0,30	1,99	0,99	0,00	9,94	0,00	0,99	6,96	10,93	31,81	58,46	18,70	32,00
H26	H	Low-energy fluvial	69,38	0,00	4,96	0,30	3,96	1,98	0,00	11,89	0,30	0,30	1,98	4,96	25,37	26,50	12,90	0,87
H27	H	Wet sandflat	56,83	1,00	6,98	1,99	5,98	0,00	0,00	6,98	0,30	1,99	2,99	14,96	33,20	56,56	19,20	22,00
H28	H	Wet sandflat	60,00	1,00	7,00	1,00	2,00	1,00	0,00	9,00	0,00	2,00	7,00	10,00	31,00	56,88	17,80	12,00
H29	H	Dune	56,32	0,30	8,89	0,30	0,30	0,30	0,00	6,92	0,00	0,99	5,93	19,76	34,19	68,54	20,50	318,00
H30	H	Dune	63,00	2,00	7,00	2,00	3,00	1,00	0,00	4,00	0,00	1,00	5,00	12,00	26,00	55,14	19,80	28,00
H31	H	Low-energy fluvial	66,21	0,30	4,94	0,59	7,91	0,00	0,00	5,93	0,30	0,99	2,96	9,88	27,96	55,70	21,50	39,00
I1	I	Damp sand flat	N/A	N/A	N/A	N/A	N/A	N/A	N/A	N/A	N/A	N/A	N/A	N/A	N/A	82,40		
I2	I	Damp sand flat	N/A	N/A	N/A	N/A	N/A	N/A	N/A	N/A	N/A	N/A	N/A	N/A	N/A	71,90		
I3	I	Dry sandflat	N/A	N/A	N/A	N/A	N/A	N/A	N/A	N/A	N/A	N/A	N/A	N/A	N/A	64,50		
I4	I	Damp sand flat	N/A	N/A	N/A	N/A	N/A	N/A	N/A	N/A	N/A	N/A	N/A	N/A	N/A	43,08		
I5	I	Dry sandflat	N/A	N/A	N/A	N/A	N/A	N/A	N/A	N/A	N/A	N/A	N/A	N/A	N/A	41,20		
J1	J	Low-energy fluvial	59,11	0,00	4,93	0,30	3,94	0,00	0,00	20,00	0,59	0,30	7,88	2,96	35,67	56,10	8,50	0,58
J2	J	Low-energy fluvial	46,31	0,30	14,78	0,30	2,96	0,30	0,00	4,93	0,30	0,30	7,88	21,67	38,33	65,60	29,60	2120,00
J3	J	Dry sandflat	64,81	1,00	1,00	1,00	2,99	0,00	0,00	19,94	0,30	1,00	3,99	3,99	32,20	58,00	10,00	0,53
J4	J	Dry sandflat	52,53	0,00	1,98	0,59	5,95	0,00	0,00	4,96	0,30	0,00	5,95	27,75	44,90	45,90	26,70	341,00
J5	J	Wet sandflat	11,93	0,30	0,99	4,97	79,52	0,00	0,00	0,30	0,00	0,00	0,00	1,99	81,81	69,40	8,50	4,80
J6	J	Low-energy fluvial	54,35	0,30	12,85	0,30	1,98	5,93	0,00	3,95	0,30	0,30	8,89	10,87	32,21	46,40	17,70	15,00
J7	J	Dune	39,53	0,00	8,89	0,59	2,96	0,30	0,00	8,89	0,99	0,30	12,85	24,70	50,99	40,50	26,20	1730,00
J8	J	Dune	51,38	0,30	7,91	0,30	0,99	0,30	0,00	9,88	0,30	1,98	8,89	17,79	40,12	77,36	24,00	544,00
J9	J	Dune	56,66	0,00	0,99	0,99	4,97	0,30	0,00	8,95	0,00	0,30	6,96	19,88	41,35	30,10	20,70	128,00
J10	J	Dune	55,67	0,30	8,95	0,30	1,99	0,99	0,00	6,96	0,00	0,99	8,95	14,91	34,79	58,40	25,50	417,00
J11	J	Dune	52,53	0,00	3,96	0,30	2,97	0,99	0,00	4,96	1,29	0,30	9,91	22,79	43,21	38,50	21,50	218,00
J12	J	Dune	53,52	0,30	9,91	0,99	4,96	0,30	0,00	6,94	0,00	0,30	5,95	16,85	35,28	45,10	26,10	596,00
J13	J	Low-energy fluvial	60,64	0,00	4,97	0,30	0,30	0,99	0,00	11,93	0,99	0,99	8,95	9,94	34,10	43,40	15,00	11,00
J14	J	Damp sandflat	55,50	0,00	3,96	0,30	0,99	0,30	0,00	5,95	0,30	0,99	11,89	19,82	40,24	34,60	25,40	293,00
J15	J	Dune	55,83	1,00	4,99	0,30	1,99	0,00	0,00	4,99	2,99	0,00	9,97	17,95	37,89	51,70	23,50	122,00

Table A.2 (continued)

Sample ID	Well	Depositional Environment	Quartz	Feldspars	Rock fragments	Accessories	Depositional matrix	Quartz cement	Feldspar cement	Carbonate cement	Sulfate cement	Illite cement	Kaolinite cement	Intergranular porosity	Intergranular volume (%)	Coat coverage (%)	Plug porosity (%)	Plug permeability (mD)
J16	J	Damp sandflat	47,43	0,00	10,87	0,30	0,99	0,30	0,00	3,26	2,27	0,99	12,85	20,75	41,40	53,68	22,70	175,00
J17	J	Damp sandflat	62,62	0,00	1,99	0,30	4,97	0,00	0,00	10,93	0,30	1,99	12,92	3,98	35,09	45,00	15,40	6,40
J18	J	Dry sandflat	54,60	0,00	3,34	1,01	6,07	0,00	0,00	12,13	0,30	0,30	13,14	9,10	41,05	38,80	10,60	0,25
J19	J	Dune	49,55	0,30	19,82	0,99	1,98	0,30	0,00	5,95	0,99	0,30	9,91	9,91	29,34	57,86	17,00	42,00
J20	J	Dune	42,62	0,00	7,93	0,30	6,94	0,00	0,00	5,95	1,29	0,30	11,89	22,79	49,16	52,98	21,20	274,00
J21	J	Dry sandflat	48,28	0,30	10,84	0,30	1,97	0,30	0,00	4,93	0,59	1,97	14,78	15,76	40,30	64,38	26,30	350,00
J22	J	Dune	48,42	0,30	9,88	0,30	1,98	0,30	0,00	5,93	1,28	0,99	9,88	16,80	41,11	61,96	20,60	84,00
J23	J	Dune	55,50	0,30	8,92	0,99	5,95	0,99	0,00	5,95	0,59	0,99	8,92	10,90	34,29	51,90	22,80	138,00
J24	J	Dune	48,37	0,30	2,96	0,00	4,94	0,00	0,00	9,87	0,00	0,99	7,90	24,68	48,37	61,30	24,30	260,00
J25	J	Dune	50,55	0,30	7,93	0,30	3,96	0,99	0,00	6,94	0,30	1,98	6,94	19,82	40,93	68,64	25,90	144,00
J26	J	Dry sandflat	46,26	0,00	2,95	0,30	5,91	0,98	0,00	8,86	0,30	0,98	8,86	24,61	50,49	66,78	23,60	40,00
J27	J	Dune	58,47	0,00	2,28	0,30	1,98	0,00	0,00	11,89	1,29	1,98	4,96	16,85	38,95	41,60	16,70	12,00
J28	J	Dune	50,40	0,30	9,88	0,30	4,94	1,98	0,00	7,91	0,59	1,98	6,92	14,82	39,13	52,10	25,80	176,00
J29	J	Dune	52,68	0,00	2,98	0,00	0,99	0,00	0,00	8,25	1,29	0,00	6,96	26,84	44,33	63,40	23,60	137,00
J30	J	Dune	48,56	0,30	7,93	0,30	4,96	0,99	0,00	6,94	1,29	0,99	10,90	16,85	42,91	47,40	26,40	197,00
J31	J	Dune	55,83	0,00	2,99	0,30	2,99	0,00	0,00	4,99	2,99	0,00	6,98	22,93	40,88	52,70		
J32	J	Dune	51,38	0,30	7,91	0,30	1,98	1,98	0,00	6,23	1,28	1,98	10,87	15,81	40,12	72,22	18,90	18,00
J33	J	Dune	48,71	0,30	9,94	0,99	3,98	1,99	0,00	6,96	3,28	0,99	6,96	15,90	40,06	78,82	28,60	428,00
J34	J	Dune	40,51	0,00	2,27	0,59	7,91	0,00	0,00	7,91	2,96	0,30	13,83	23,72	56,62	53,20	29,40	394,00
J35	J	Dune	56,83	0,00	7,98	0,00	2,99	0,30	0,00	3,99	2,99	0,00	6,98	17,95	35,19	58,10	18,10	18,00
J36	J	Low-energy fluvial	51,69	2,98	5,96	2,98	3,98	4,97	0,00	6,96	2,29	0,30	3,98	13,92	36,38	67,48	18,00	2,50
J37	J	Low-energy fluvial	62,01	0,00	2,26	0,30	4,92	0,98	0,00	13,78	1,97	0,98	7,87	4,92	35,43	44,68	7,50	0,65
J38	J	Dry sandflat	62,25	0,00	2,96	0,30	1,98	0,30	0,00	9,88	0,30	0,30	3,95	17,79	34,49	56,20	20,70	23,00
J39	J	Dune	48,85	1,00	6,98	1,99	8,97	1,99	0,00	6,98	1,30	1,00	5,98	14,96	41,18	53,30	26,10	229,00
J40	J	Low-energy fluvial	63,81	1,00	2,99	0,30	2,99	1,99	0,00	6,98	4,99	1,00	6,98	6,98	31,90	47,30	15,40	13,00
J41	J	Damp sandflat	57,83	2,99	1,00	0,30	11,96	1,99	0,00	9,97	3,99	1,99	4,99	2,99	37,89	61,80		

Table A.3: Compilation of point count data and sample information of Bebertal samples.

Sample ID	Depositional Environment	Quartz	Feldspars	Rock fragments	Accessories	Iron oxides	Matrix	Quartz cement	Feldspar cement	Carbonate cement	Carbonate replacement	Sulfate cement	Illite cement	Illite replacement	Iron oxide cement	Intergranular porosity	Secondary porosity	Intergranular volume (%)	Coat coverage (%)	Plug porosity (%)	Plug permeability (mD)
BEB_1	Dune	45,95	8,74	16,18	0,00	1,29	0,00	7,77	2,59	1,62	0,65	0,00	9,06	1,94	3,88	0,00	0,32	24,92	44,90	7,80	0,14
BEB_2	Dune	42,05	9,27	17,22	0,66	0,66	0,00	8,61	1,99	4,30	0,33	0,00	10,26	1,66	2,65	0,00	0,33	27,81	40,56		
BEB_4.1	Low-energy fluvial	51,41	4,08	15,05	0,00	0,94	0,00	7,52	0,00	4,39	0,00	0,00	7,21	1,57	7,84	0,00	0,00	26,96	60,22		
BEB_4.2	Low-energy fluvial	43,00	4,00	37,00	0,33	0,33	0,00	2,67	1,33	0,00	0,00	0,00	4,00	3,00	4,33	0,00	0,00	12,33	60,46		
BEB_4.3	Low-energy fluvial	50,33	4,97	7,28	0,00	1,66	0,99	3,97	0,00	0,33	0,00	0,00	11,26	4,30	14,90	0,00	0,00	31,46	56,22		
BEB_5	Dune	45,00	6,00	14,00	0,33	0,00	0,00	11,67	0,67	0,33	0,00	0,00	9,00	1,33	6,67	1,33	3,67	29,67	56,90	13,36	115,24
BEB_6	Dune	43,00	6,00	15,33	1,67	1,33	0,00	7,33	1,00	4,00	0,00	0,00	12,67	2,33	5,00	0,33	0,00	30,33	50,62		
BEB_7	Dune	42,67	5,67	20,33	0,33	1,00	0,00	4,67	1,67	3,00	0,00	0,00	12,33	2,33	6,00	0,00	0,00	27,67	53,58	8,59	110,61
BEB_8	Low-energy fluvial	33,33	3,67	42,67	0,00	1,33	0,67	4,33	0,33	0,00	0,00	0,00	11,00	0,00	2,67	0,00	0,00	19,00	62,86	3,80	4,13
BEB_8_C	Low-energy fluvial	N/A	N/A	N/A	N/A	N/A	N/A	N/A	N/A	N/A	N/A	N/A	N/A	N/A	N/A	N/A	N/A	N/A	N/A	3,88	0,58
BEB_D1	Low-energy fluvial	54,33	5,33	12,67	0,00	1,33	0,00	6,00	0,67	0,00	0,00	0,00	10,33	1,00	8,00	0,00	0,33	25,00	62,82		
BEB_D2	Dune	41,58	6,60	21,12	0,00	0,66	0,00	6,60	2,31	0,66	0,00	0,33	8,91	1,65	7,59	0,99	1,32	27,06	42,86	9,79	3,59
BEB_D3.1	Dune	49,33	4,00	14,33	0,67	1,00	0,00	11,00	2,00	0,00	0,00	0,67	11,67	1,33	4,00	0,00	0,00	29,33	47,90		
BEB_D3.2	Dune	53,20	6,69	12,79	0,00	0,87	0,00	9,30	0,00	0,00	0,00	0,00	9,59	2,03	5,23	0,00	0,29	24,13	45,74		
BEB_D3.3	Dune	46,86	3,30	14,19	0,33	0,99	0,33	6,60	0,00	0,00	0,00	0,00	12,54	2,97	11,88	0,00	0,00	31,35	49,86		
B1	Dune	38,67	6,00	23,00	0,00	1,00	1,00	14,00	2,33	0,00	0,00	0,33	7,00	3,67	2,67	0,00	0,33	27,33	25,70		
B2	Dune	38,00	13,00	13,00	0,00	0,67	0,33	16,33	2,67	2,00	0,67	0,00	4,67	2,67	3,33	1,67	1,00	31,00	30,72		
B9	Dry Sandflat	38,67	12,67	7,67	0,00	1,00	1,33	9,00	2,67	1,00	1,00	0,00	12,67	5,67	6,00	0,00	0,67	32,67	50,32		
BEBC_1	Dune	38,33	6,67	18,67	0,33	1,00	0,67	6,67	2,67	0,00	0,00	0,00	8,33	3,33	3,67	5,67	4,00	27,67	42,24		
SQ2_1-1	Dune	44,00	6,00	15,33	0,00	1,33	0,33	4,00	2,00	7,00	0,00	0,00	9,33	2,00	3,67	1,67	3,33	28,00	47,80	10,83	0,31
SQ2_1-2	Dune	35,33	4,67	19,67	0,00	2,00	1,33	4,00	1,00	16,67	3,33	0,00	2,67	1,67	5,00	0,00	2,67	30,67	31,74	7,12	0,21
SQ2_1-3	Dune	35,00	7,00	17,33	0,00	0,67	1,33	9,67	3,00	8,00	1,33	0,00	7,00	1,33	2,67	2,67	3,00	34,33	38,98	12,62	0,40
SQ2_4-2	Dune	40,33	10,00	8,67	0,00	1,67	1,67	7,67	4,00	2,33	0,00	0,00	7,00	3,33	2,67	4,00	6,67	29,33	34,72	13,04	0,36
SQ2_4-3	Dune	33,33	8,33	16,00	0,00	1,00	4,33	5,00	2,00	9,67	2,33	0,00	5,00	2,67	4,33	2,00	4,00	32,33	32,42	9,47	0,17
SQ2_8-1	Dune	41,00	10,33	13,00	0,33	1,33	1,33	5,00	2,33	2,00	0,00	0,00	12,00	1,33	2,00	4,67	3,33	29,33	49,24	16,07	3,14
SQ2_8-2	Dune	38,33	7,67	17,00	0,00	0,67	1,67	10,33	3,00	2,67	0,00	0,00	8,33	1,33	2,67	2,33	4,00	31,00	30,12	12,94	1,07
SQ2_8-3	Dune	45,33	5,00	8,00	0,33	1,33	1,00	8,67	4,33	2,00	0,67	0,00	10,00	2,33	1,67	4,67	4,67	32,33	32,54	14,58	0,53
SQ3_1-1	Dune	52,67	5,67	11,00	0,33	0,67	2,00	3,67	2,33	2,67	0,00	0,67	13,33	2,00	0,33	1,67	1,00	26,67	49,26	12,29	0,38
SQ3_1-2	Dune	46,33	9,67	11,33	0,00	1,33	1,33	5,00	3,33	0,33	0,00	0,00	11,33	1,67	2,00	4,33	2,00	27,67	48,42	11,97	0,15
SQ3_1-3	Dune	42,67	9,00	9,00	0,00	1,00	1,00	5,00	3,00	0,33	0,00	0,00	13,67	3,33	4,67	4,33	3,00	32,00	45,32	13,58	0,17
SQ4_1-2	Dune	41,33	4,00	19,33	0,00	1,67	1,00	1,33	1,00	6,33	0,33	1,33	7,67	3,00	2,67	5,67	3,33	27,00	32,28	12,83	0,24
SQ4_1-3	Dune	40,33	7,00	18,67	0,33	1,00	1,67	4,67	3,67	8,00	0,67	0,00	8,67	2,33	1,00	1,00	1,00	28,67	31,70	10,54	0,19
SQ4_3-2	Dune	44,00	6,67	21,00	0,00	0,00	2,67	2,67	0,67	3,33	0,67	2,33	7,00	1,00	3,33	2,33	2,33	24,33	38,02	11,58	0,21
SQ4_3-3	Dune	42,67	8,33	18,00	0,00	1,00	2,00	3,33	2,00	4,33	0,33	2,33	6,00	1,67	0,67	4,33	3,00	25,00	29,60	10,43	0,60

Table A.3 (continued)

Sample ID	Depositional Environment	Quartz	Feldspars	Rock fragments	Accessories	Iron oxides	Matrix	Quartz cement	Feldspar cement	Carbonate cement	Carbonate replacement	Sulfate cement	Illite cement	Illite replacement	Iron oxide cement	Intergranular porosity	Secondary porosity	Intergranular volume (%)	Coat coverage (%)	Plug porosity (%)	Plug permeability (mD)
SQ4_4-1	Dune	42,00	8,00	14,00	0,00	1,33	3,00	2,33	2,33	0,00	0,00	0,33	12,33	2,67	3,67	5,67	2,33	29,67	44,14	14,00	0,21
SQ4_4-2	Dune	43,00	8,67	16,33	0,00	1,00	2,00	3,00	1,67	4,00	0,00	0,67	8,33	2,33	3,67	3,33	2,00	26,67	46,80	12,51	0,09
SQ4_4-3	Dune	38,67	7,67	23,00	0,00	0,33	3,33	4,67	1,00	3,00	0,67	2,00	3,33	5,33	1,33	2,67	3,00	21,33	37,96	16,27	2,53

Table A.4: Compilation of point count data and sample information of N-England samples from the Vale of Eden.

Sample ID	Depositional Environment	Quartz	Feldspars	Rock fragments	Iron oxide	Depositional matrix	Accessories	Quartz cement	Feldspar cement	Carbonate cement	Carbonate replacement	Sulfate cement	Illite cement	Illite replacement	Kaolinite cement	Kaolinite replacement	Iron oxide cement	Secondary porosity	Intergranular porosity	Intergranular volume (%)	Coat coverage (%)	Plug porosity (%)
BS_1	Fluvial channel	70,67	4,00	0,67	0,00	4,00	0,00	0,33	0,00	0,33	0,00	0,00	5,00	3,67	0,00	0,00	0,33	0,00	11,00	21,00	70,06	
BS_2	Fluvial channel	69,33	5,67	2,67	2,00	2,00	0,67	0,33	0,00	2,67	0,00	0,00	2,67	3,00	0,00	0,00	1,33	0,33	7,33	16,33	79,9	
BS_congl	Alluvial fan	64,00	1,67	11,33	0,33	1,33	0,33	0,00	0,00	14,33	3,00	0,00	1,00	0,67	0,00	0,00	0,00	0,33	1,67	18,33	n.a.	
BQ_1	Dune	68,33	2,67	0,33	0,00	0,00	0,00	20,67	0,33	0,00	0,00	0,00	0,00	1,00	1,33	0,00	0,00	0,33	5,00	27,33	13,6	
BQ_2	Dune	64,67	6,33	1,67	0,00	0,00	0,00	11,00	0,00	0,00	0,00	0,00	1,67	1,67	1,33	0,00	1,67	2,00	8,00	23,67	25,52	
BQ_3	Dune	71,33	5,00	0,67	0,00	0,33	0,00	11,33	0,00	0,00	0,00	0,00	0,33	0,00	1,33	0,00	0,33	0,67	8,67	22,33	13,9	
BQ_4	Dune	60,00	6,33	2,33	0,00	0,00	0,00	18,00	0,00	0,00	0,00	0,00	0,33	1,67	0,67	0,00	1,33	0,00	9,33	29,67	21,16	
FM_1	Dune	78,67	4,33	1,33	0,00	0,33	0,00	1,00	0,00	0,00	0,00	0,00	2,33	0,67	0,33	0,00	0,67	1,00	9,33	14,00	95,32	
FM_2	Dune	67,00	6,33	3,00	0,67	0,00	0,00	11,67	0,33	0,00	0,00	0,00	1,33	2,67	0,00	0,00	2,00	0,00	4,67	20,33	37,12	
FM_3	Dune	65,67	4,33	1,67	0,67	0,00	0,00	7,00	0,67	0,00	0,00	0,00	1,67	2,00	0,33	0,00	1,67	0,67	13,67	25,00	61,74	
FM_Alt	Dune	72,09	2,99	1,33	0,00	0,00	1,00	2,33	0,00	0,66	0,33	0,33	7,97	1,00	0,66	0,00	0,33	0,33	8,64	20,93	83,88	
LC_1	Dune	71,33	4,67	2,00	0,33	0,00	0,33	5,00	0,00	0,00	0,00	0,00	2,67	2,33	0,33	0,00	0,67	0,67	9,67	18,33	70,06	
LC_2	Dune	76,00	2,67	1,33	0,00	0,00	0,00	7,33	1,00	0,00	0,00	0,00	2,67	2,67	0,00	0,00	0,33	0,67	5,00	16,67	74,48	
SH_1	Dune	63,70	4,29	3,30	0,00	0,33	0,00	16,50	0,33	0,00	0,00	0,00	2,31	2,31	1,32	0,00	0,33	0,66	4,62	25,74	19,1	
RR_1	Dune	65,00	4,67	2,33	0,00	0,00	0,00	13,33	0,00	0,00	0,00	0,00	2,00	4,33	1,33	0,00	1,67	1,67	3,67	22,00	31,18	
RR_2	Dune	66,33	6,67	1,00	0,00	0,00	0,33	14,67	0,00	0,00	0,00	0,00	2,00	4,33	1,00	0,00	0,33	1,67	1,67	19,67	20,3	
RR_3	Dune	73,00	2,67	2,00	0,00	0,00	0,00	13,67	0,67	0,00	0,00	0,00	0,33	1,67	1,00	0,00	0,67	0,33	4,00	20,33	15,18	
RR_4	Dune	69,33	4,33	2,33	0,00	0,00	0,67	12,33	0,33	0,00	0,00	0,00	0,67	2,33	0,67	0,00	0,00	0,00	7,00	21,00	19,8	
RR_5	Dune	70,67	4,67	1,67	0,33	0,33	0,33	5,67	0,00	0,00	0,00	0,33	0,67	3,33	0,67	0,00	2,00	1,33	8,00	17,67	38,18	
BQ_6_2-1	Dune	61,00	2,00	2,00	0,33	0,00	0,00	16,00	0,00	0,00	0,00	0,00	1,33	0,00	0,00	0,00	7,33	1,33	8,67	33,33	6,74	18,13
BQ_6_2-2	Dune	57,33	2,67	1,67	0,00	0,00	0,00	21,33	0,00	0,00	0,00	0,00	1,00	0,33	0,00	0,33	2,33	2,33	10,67	35,33	7,54	19,42
BQ_6_2-3	Dune	69,00	2,00	1,33	0,00	0,00	0,00	16,33	0,00	0,00	0,00	0,00	0,33	0,33	0,00	0,33	2,67	1,67	6,00	25,33	5,8	17,55
BQ_6_3-1	Dune	67,33	3,00	2,67	0,00	0,00	0,00	18,33	0,00	0,00	0,00	0,00	0,00	0,00	0,00	0,67	3,33	2,33	2,33	24,00	4,8	20,32
BQ_6_3-2	Dune	67,33	1,67	0,33	0,00	0,00	0,00	20,00	0,00	0,00	0,00	0,00	1,33	0,33	0,00	0,00	3,00	4,33	1,67	26,00	5,08	19,94
BQ_6_3-3	Dune	65,67	2,33	1,33	0,00	0,00	0,00	22,33	0,00	0,00	0,00	0,00	0,00	0,00	0,00	0,67	2,00	1,33	4,33	28,67	4,14	19,99
BQ_6_5-1	Dune	69,33	2,00	0,33	0,00	0,00	0,00	15,00	0,00	0,00	0,00	0,00	0,67	1,00	0,67	0,33	2,33	3,33	5,00	23,67	4,52	20,33
BQ_6_5-2	Dune	67,00	1,00	1,33	0,00	0,00	0,00	19,00	0,00	0,00	0,00	0,00	0,00	0,67	0,00	0,67	3,00	1,00	6,33	28,33	5,12	19,52
BQ_6_5-3	Dune	65,33	2,00	2,00	0,00	0,00	0,00	10,00	0,00	0,00	0,00	0,00	0,00	1,00	0,67	0,33	4,07	2,33	3,07	27,00	6,96	18,74